# PLASTIC COMPOSITES FOR 21st CENTURY CONSTRUCTION

Proceedings of a session sponsored by the
Materials Engineering Division of the
American Society of Civil Engineers in conjunction with
the ASCE Convention in Dallas, Texas,
October 24-28, 1993

**Edited by Richard E. Chambers**

Published by the
American Society of Civil Engineers
345 East 47th Street
New York, NY 10017-2398

## ABSTRACT

This proceedings, *Plastics Composites for 21st Century Construction,* consists of papers presented at a session of the 1993 ASCE Annual Convention and Exposition held in Dallas, Texas October 24-28, 1993. These papers explore both the current status and the developments underway to introduce composites as structural materials in the infrastructure. The topics covered concern the market development of composites, composite pultruded structural shapes, reinforcing bars and grids and tendons for concrete, and standards development for composites.

Library of Congress Cataloging-in-Publication Data

Plastics composites for 21st century construction: proceedings of a session/sponsored by the Materials Engineering Division of the American Society of Civil Engineers in con junction with the ASCE Convention in Dallas, Texas, October 24-28, 1993;edited by Richard E. Chambers.
    p. cm.
Includes index.
ISBN 0-87262-989-9
    1.Composite materials—Congresses. 2.Plastics—Congresses. I.Chambers, R.E. (Richard E.) II. American Society of Civil Engineers. Materials Engineering Division. III.ASCE    National Convention (1993:Dallas, Tex.)
TA418.9.C6P545 1993
624.1'8923—dc20                        93-32978
                                        CIP

The Society is not responsible for any statements made or opinions expressed in its publications.

Photocopies. Authorization to photocopy material for internal or personal use under circumstances not falling within the fair use provisions of the Copyright Act is granted by ASCE to libraries and other users registered with the Copyright Clearance Center (CCC) Transactional Reporting Service, provided that the base fee of $2.00 per article plus $.25 per page copied is paid directly to CCC, 27 Congress Street, Salem, MA 01970. The identification for ASCE Books is 0-87262-989-9/93 $2.00 + $.25. Requests for special permission or bulk copying should be addressed to Permissions & Copyright Dept., ASCE.

Copyright © 1993 by the American Society of Civil Engineers,
All Rights Reserved.
Library of Congress Catalog Card No: 93-32978
ISBN 0-87262-989-9
Manufactured in the United States of America.

Cover photo courtesy of the Advanced Composite Technology Consortium (host: UCSD).

# PREFACE

Given my assignment by the Structural Composites and Plastics Committee of the Materials Engineering Division to create a technical session on composites for this 1993 ASCE Annual Convention & Exposition it turned out to have a highly appropriate theme—*Moving into the 21st Century*. It seemed like a very opportune occasion to present attendees a picture of both the current status and the developements underway to introduce composites as structural materials in the infrastructure. Indeed the timing couldn't be better because so much is happening so fast. In the last several months alone, the Civil Engineering Research Foundation (CERF) has become to recommend strategies at a National policy level on innovative materials and applications of composites in infrastructure, ASCE approved an effort to write design and construction standards for composites, and FHWA has announced that it is contributing to a project to construct a full-scale cable-stay bridge at a site at the University of California at San Diego!

I have asked three leaders in the industry to develop a capsule summary of his area of expertise—Douglas S. Barno with long experience in the market development of composites representing the visions and goals of the Composites Institute, Dr. Ayman S. Mosallam of George Washington University on composite pultruded structural shapes, Drs. Salem S. Faza and Hota V. S. GangaRao of the Constructed Facilities Center of West Virginia University on reinforcing bars and grids and tendons for concrete. Having been involved in the ASCE initiative on standards development for composites, I offer some concluding remarks and a perspective on that subject. We hope that the civil engineering practitioner and those in the composites industry who are increasingly involved in civil engineering applications find this endeavor enlightening.

The papers were requested, reviewed and accepted by me and my paper received a peer review as well. All papers are eligible for discussion in the *Journal of Materials in Civil Engineering*. All papers are also eligible for ASCE awards.

Richard E. Chambers

# CONTENTS

## Session 21
## Plastics Composites for 21st Century Construction
## Moderators: Richard E. Chambers & Ahmed Morsi

The Use of Structural Composites in 21st Century Infrastructure Technology
  Douglas S. Barno .................................................. 1
Composite Materials as Concrete Reinforcement in the Next Decade
  Salem S. Faza and Hota V. S. GangaRao ............................. 15
Pultruded Composites: Materials for the 21st Century
  A. S. Mosallam ................................................... 23
21st Century Composites Require ASCE Standards and SDS
  Richard E. Chambers .............................................. 56

**Subject Index** .................................................... 73

**Author Index** ..................................................... 74

# THE USE OF STRUCTURAL COMPOSITES IN 21st CENTURY INFRASTRUCTURE TECHNOLOGY

## ABSTRACT

The current interest in composites as materials of construction for structural applications in the 21st century infrastructure is driven in large part by the recognition that composites offer certain inherent advantages over traditional materials. While it is apparent that fiber/polymer composites by themselves will play an increasingly important role in the future of the infrastructure, there is even greater promise for the new concept of joining composites with traditional materials (wood, concrete and steel) to form "hybrids" or "super-composites."

The advantages of all-composite structures and hybrids are more widely understood and appreciated in industrial/original-equipment manufacture (OEM) environments than in civil engineering. The purpose of this paper is to create and transmit a sense of the unique benefits which effective application of composites technology can bring to the civil engineering infrastructure arena. Some of these benefits can include:

- stronger, more efficient structures which may still employ traditional materials

- new combinations of materials whose total performance exceeds the limits of performance of either material(s) by itself.

- composites which take full advantage of the properties of each constitutent material

- the ability to tailor properties that are complimentary to traditional materials is perhaps one of the most appealing benefits of composites


Author:
Douglas S. Barno, Consultant to:
The Composites Institute of the Society of The Plastics Industry, Inc.
DSB Associates
2635 Old Columbus Rd.
Granville, OH  43023

## NEW COMPOSITES PHILOSOPHY

Composites may not be required in many applications. However, when traditional materials alone are unsuited, composite structures and hybrids of composites plus traditional materials may expand the performance envelope. Historically, composites and traditional materials have been mutually exclusive. The trick will be to translate the benefits of composites into new designs and new combinations of materials to achieve civil engineering structures that will outperform competitive products on the basis of cost, performance and productivity.

In this paper, we are articulating, for the first time, a new "composite philosophy" in which individual materials (including steel, concrete, timber, etc.) are joined in new and unique combinations that offer improved performance compared to each material by itself. This combination of materials is powerful. Each material is used where its particular cost and mechanical/physical/durability performance story is "unbeatable." The philosophy being articulated is not that of a specific material technology, but, rather that of a process.

For example, at the University of West Virginia, the load-bearing performance of wood beams is being improved by over 30% by adhering a <1/4" thick composite strip on the tensile face of the wooden beam. This marriage of composites and wood results in better performance than either could provide by itself. This is, in fact, a logical extension of the very efficient "composite" construction developed by structural engineers for bridges, which ties concrete decks to steel girders to provide composite structural action.

In order to effectively apply this new composites philosophy, it will be important to address the following critical issues:

- identifying specific applications where composites can bring the greatest value and most performance-additive for development and demonstration in the rebuilding of America.

- composites should not necessarily be viewed as a binary allternative to iron, steel, concrete, wood, etc. In many cases, it is possible to greatly increase the performance of existing structures by new and unique combinations of composites and traditional materials mentioned above.

- much of the development and technology transfer envisioned in this report is based on using composites by themselves or in combination with traditional materials to improve performance in existing applications. This is a form of technology substitution.

- composites with their unique properties can allow infrastructure structures that cannot be undertaken with traditional materials alone. An example is a composite bridge over the Straits of Gibralter. While theoretically possible with composites, such a structure could not be built in steel or concrete as the weight of the bridge would exceed the load carrying capabilities of either steel or concrete. In cases such as this, composites would become an enabling technology.

## A LOOK AT THE FUTURE OF ADVANCED CONSTRUCTION TECHNOLOGY

There is potential to use composites to improve infrastructure performance and productivity in each of the short, medium and long terms. Conventional wisdom suggests that investigations should start with state-of-the-art technology, vectoring forward using today's knowledge base as the point-of-departure.

The problem with this approach is that it may unwittingly place boundaries on creativity and innovation by defining the future in relation to the present and past. Later in this paper, the issues of excessive constriction of the innovation process and its potential adverse effects on composite infrastructure development are addressed.

As the result of aggressive implementation of the strategy advanced here, we believe that the following industry-shaping developments *can be in common usage throughout the U.S. construction industry by the year 2000!*

- Structural stay-in-place formwork for concrete construction
- Composite prestressing tendons and chucks for prestressed columns, beams and slabs
- "Super-Prestressing" of concrete structurals using composites
- Optimized/tailored/standardized modular bridge decks
- Composite structural piling and sheet pile for waterfront and inland installations
- Magnetic Levitation (MagLev) infrastructure and right-of-way using composites
- Seismic-resistant structures
    - new construction
    - upgrading of existing structures
- Modular, standardized composite structural components
    - piers
    - bridges
    - slabs
    - retaining walls
    - earth anchors
    - deadmen tie-backs
    - dams, locks and navigation aids
- High rise buildings
    -structurals
    - claddings/skins
- Piping
    - potable water
    - sewage
    - storm drains
    - irrigation
- Tunnels

- Comprehensive Composites Design Protocol
    - Total Structural Knowledge (TSK) of composite materials behavior and a structural design system (SDS) for selected applications
        - data base
        - design equations
        - composites design handbook
        - design protocols on interactive CD ROM disks
        - ease of design (in this case, developments will make composite materials more "designer/user-friendly")
        - standardized practices, codes and specifications
    - Long term testing and accelerated aging test methods with high correlation to actual long-term performance
- National standards and codes for composite materials and structures that facilitate continuing materials, design and construction innovation
- U.S. and worldwide technical & installations clearing house
    - data base
    - installations/case histories
    - ongoing research
- Practitioner education programs
    - undergraduate engineering curriculum
    - practitioner continuing education
    - certified composite designer program
    - certified fabricator program
- Composite stacks & stack liners
- Industry-accepted products
    - concrete reinforcing systems
        - dowel rods
        - reinforcing rods, grids and 3-D structural network
        - prestressing and post-tensioning tendons
    - suspension cables (cable stays)
    - beams and structural shapes optimized for geometry and performance
    - cladding, skins and curtain walls
    - decking (bridges & piers)
    - piling, bulkheads and sheet pile
    - formwork (reuseable and stay-in-place)
    - modular construction
- Hybrid structurals combining composites and conventional materials
- On-site fabrication technology (reinforcing rods, grids, pipe, structurals, etc.)
- 100-year design life structures based on composites & hybrids
- Infrastructure rehabilitation systems
    - standardized materials & construction practices
    - column wrapping (corrosion & seismic considerations)
    - soffit plates
    - structural cladding
    - in-situ repair of civil engineering structures
    - in-situ structural or appearance upgrading of infrastructures

- Signs & signposts
- Highway guardrails & posts
- Lighting standards
- Utility poles
- Railroad ties
- "Smart structures" (self-diagnosing, early warning & detection of failure)
- Crashworthy highway/roadside appurtenances
- Integrated modular structural stay-in-place formwork
- Certified design/build fabricators
- High Speed Rail (right-of-way, security, rolling stock, utilities, etc.)
- Environmentally safe electrical transmission towers
- Desalination plants
- National Coordinating Council Civil Engineering Center
    - data
    - testing
    - design
    - materials selection
    - training
    - education
    - demonstration jobs
    - research
    - technology transfer
    - practitioner education
    - liason to civil engineering/infrastructure community
- Optimized composites materials systems
    - high elongation, environment-resistant resins (alkali)
    - high shear strength matrix
    - non-brittle (ductile) failure mechanisms
- Connectivity
    - composite to composite
    - composite to traditional materials
    - engineered standardized connection modules
- Fire safety (flame & smoke)
- Recycleability
    - recycle composite structural components
    - component reuse
    - incorporate post-consumer recycled materials into composites
- Modular water and wastewater treatment systems
    - large diameter tanks
    - process piping
    - ancillary equipment (railings, grating, wiers, diffusers, etc.)
    - pollution control covers
- Environmentally friendly materials systems
    - don't break down in service
    - cross-linked polymers inherently resist deterioriation
    - unaffected by even harsh environments

- Offshore civil engineering applications
  - oil production platforms
    - housing units
    - riser pipe
    - platform structures
    - cables
    - safety equipment
    - tension-leg platform legs
  - cargo transfer stations
  - floating cities
  - marine/maritime industry structures
  - mariculture
- Dams & Locks
  -water control mechanisms
  - navigation aids (bouys, etc.)

The importance of "vision" in achieving these goals cannot be overstated. Bartlett's Famous Quotations notes that, "It is the commonest of errors to believe that the limits of our perception are the limits of all that there is to perceive."

In their current bestselling book, "Re-Engineering the Corporation, a Manifesto for Business Revolution" by Michael Hammer and James Champy published by Harper Business, the authors point out that current organizations are not well suited to create change or prepare for the future. Within the boundaries of existing organizations and industries, aspirations and needs are shaped by what is believed to be possible. Therefore, failure to innovate or "leap-frog" forward using technology is largely the result of a failure to conceive and impart alternative visions of what may be possible. In an example from the book, IBM turned down an attractive offer to take over Xeroxgraphy in the early 1950's because this technology was seen *only as an alternative to carbon paper!* The enormous copier industry which has grown throughout the world was beyond the vision of the evaluation team because they defined the potential application by the then-dominant method of multiple-copy copying technology... carbon paper.

### Development and Technology Transfer Required

Today, there is widespread concern that construction practices and construction technology in the U.S. lags behind the state-of-the-art in other parts of the world, particularly Japan and Western Europe. At the same time, research in the field of composite infrastructure structures in Japan and Western Europe is far advanced when compared to the U.S. The future of the U.S heavy construction industry and the future of infrastructure revitaliztion will depend in large degree on how well U.S. firms are able to catch up to and surpass prospective foreign competitors with advanced technology.

STRUCTURAL COMPOSITES 7

The composites industry has been aware of the infrastructure opportunity for several years. Initial interest was focused on those segments of the civil engineering industry where traditional materials are most noticably deficient and where the needs for enhanced performance are greatest. The marine/waterfront segment of the civil engineering infrastructure has been identified as one of the most important targets. Traditional materials (steel, concrete and wood) are inherently unsuited for long-term exposure in the harsh operating environment of marine installations. Steel is prone to corrosion and conventional practices (sandblasting, lead-based primers and solvent-based paints) are more and more in environmental disfavor.

## Conceptual Approach

The model for the ideas contained in this report arises from the recognition that the composite must be tailored for the specific applications. Parameters of tailoring should comprise not only mechanical & physical loads, chemical resistance, parts consolidation, etc., but add an evaluation for construction considerations (on-site labor savings, etc.). This leads to 3-D multi-dimensional reinforcement sub-structure or cage, consolidated in the manufacturing. Rather than continuing the conventional practice of tying rebar in the field, specialized, computer-controlled 2-D or 3-D reinforcements could be produced in a factory under controlled conditions.

Presently, the construction industry accomplishes connection pieces between beams & columns by using clip angles, plates & channels, etc. Composite technology could offer a breakthrough in jobsite savings by replacing all this jobsite labor and multiple parts required by providing a tailored molded structural connection into which the adjacent structurals are inserted.

## The Parts Consolidation Story

One of the conceptual advantages of composites for infrastructure/construction applications is the inherent ability of this versatile material family of materials to consolidate parts in singular moldings. With composites, this is a singular benefit which should not be underestimated as a technology to facilitate change in infrastructure design and construction practice.

Entire industries -most notably the automobile industry- have found that composites can provide extraordinary versatility in combining what would normally be complex, multiple-part assemblies in one trouble-free composite molding. An excellent example of the ability to benefit from parts consolidation is a typical composite automotive front end assembly. In this case, a one-piece composite molding can replace up to 22-25 individual metallic parts. Not only does the one piece composite front end reduce the number of parts by over 2,000%, the final assembly is more dimensionally accurate, lower weight by approximately 60% and lower in overall cost by approximately 50%.

Parts consolidation is a compelling story. But, in order to succeed in the construction of our nation's infrastructure, composite parts consolidation technology as well as the ability to join such structures must be made to be field or job-site "friendly." Factory moldings, produced under controlled conditions produces structures with predicable performance characteristics.

However, the field can be a different story. Field-fabricated large-diameter tanks over 65 ft. in diameter, huge stack liners up to 1,200 feet in length and other composite products are commonplace in the industrial market. In order to fully develop the potential for on-site fabrication technology, field testing and materials systems optimized for field operating conditions must be developed.

At the same time, research should be undertaken to develop a new generation of composite materials designed for infrastructure construction applications fabricated under field conditions.

Another example of unique parts consolidation will be the development of so-called, "Stay-in-PlaceStructural Formwork." Presently the civil engineering industry uses this powerful concept with corrugated steel decking in bridge slab construction. In this case, the corrugated steel functions as a sacrificial pouring form... laid in place to contain the concrete until it is cured, but making no significant long-term structural contribution to the performance of the structure. In the case of a properly designed composite stay-in-place form, the composite would remain in place to function as a permanent structural member, adding high levels of tensile strength to the tensile face of the bridge deck, just where such tensile strength is required to optimize the performance of the structure.

Bridge decking should be an ideal target for parts consolidation and weight savings. Industry experts have estimated that stay-in-place composite bridge decks may produce over 30% weight savings compared to conventional bridge decks. This means that 30% less material will be required to construct a bridge capable of carrying the same load. There is another advantage to the concept of a composite bridge deck in that there are discrete design and construction elements which can be evaluated separately to optimize the performance of the structure.

The structural steel industry provides excellent examples of how changing the paradigm of traditional construction practice can result in extraordinary corollary benefits. The Baugries organization in California has developed a new technology to address the problem of fabricating heavy wire stirrups required for elongated steel structures such as columns, piling, etc. In this case, rigid factory-welded heavy gauge steel cages replace field fabricated hooked stirups resulting in a 20% reduction of longitudinal steel bars required for a structure of equivalent strength. The steel industry is developing new forms of pre-welded multi-dimensional reinforcements. However efficient these new steel structures may be, however, composite reinforcing structures will continue to have a performance advantage in terms of inherent corrosion resistance, high tensile strength and greater strength-to-weight ratios.

**Changing the U.S. Development Paradigm**

It is a given that conventional construction is based on traditional practice. The greatest resource of the industry is its collective body of knowledge including data, codes, specs/standards, experience, etc. It is also a given that most traditional materials (wood, reinforced concrete, etc.) are composites. And arguably, the most important property of these traditional materials, which they all share, is a high level of compressive strength.

Buckminster Fuller, visionary architect/engineer and creator of the "Geodesic Dome" stated as early as the first half of the 20th century that the "next breakthrough in construction technology would be when structures move from designs based on compression to designs based materials and construction practice of tension structures. He used as easily understood examples, naturally-occuring structures such as trees. In the case of trees, complex structures based primarily on cellulose fibers operate in tension to carry incredible loads. When you consider the nature of these loads (consider a large limb 40 to 50 ft. long, loaded with leaves or snow and ice in high winds) and their relative size and durability, it is clear that there are still many lessons to be learned from natural structures.

Fuller further stated that the "state-of-the-art of conventional materials is incapable of approaching the performance of naturally-occurring structures, unless we begin to adopt some of nature's technologies." Since many natural structures are by definition composites (for example: a tree limb employs cellulose fibers for tensile strength and incompressible sap [enhanced water] for compressive strength), the comparisons are obvious. We need to use the right materials systems in the right ways and take many of our cues from nature. Consider a simple natural structure such as a cat-tail. A slender fiber/fluid column supports a weighty upper seed pod through a wide range of wind, thermal and climatic conditions while the lower part of the structure is immersed in a relatively harsh operating medium. While seemingly "fuzzy", these examples point to a time-tested model for materials and structural development which today's researchers and engineers would do well to consider.

## IMPLEMENTATION ISSUES

First generation implementation issues will concern identifying those types of existing civil engineering structures which can be restored or upgraded by incorporating innovative, but existing and proven, composite technology. Take, for example, the subject of composite bridge decking. There are several primary approaches under investigation. In one approach, an all-composite bridge deck is envisioned. In another, it is possible to take an existing deteriorated bridge, remove the failing steel and concrete decking and replace it with a new composite-reinforced prestressed concrete deck. Utilize the thinner concrete thickness achieved by prestressing with corrosion-resistant composite tendons (which will inherently resist corrosion due to de-icing salts to reduce bridge dead weight, thereby allowing higher traffic loads on the bridge. The same bridge, by virtue of selecting corrosion-resistant composite reinforcements plus application of advanced concrete technology (such as polymer concrete) should result in a much more durable structure, providing improved low-maintenance long term performance.

In concept the capability to use composites to rehabilitate or upgrade existing civil engineering structures is analogous to the practice of rehabiltating aircraft airframes in which it is possible to take an older design airframe, replace the aged piston engine with a new turboprop engine, upgrade the avionics, re-fit the interior to create an essentially new aircraft on the old structural platform. This type of rehabilitation/upgrading technology as applied to existing civil engineering structures could be an important future.

At the same time, new composite civil engineering solutions based on off-the-self or easily optimized and demonstrated product applications must be pursued. There are a number of existing products (beams, box sections, grating, handrails, etc.) which are widely used in industrial corrosion-resistant service including oil & gas production, water/wastewater treatment and chemical processing [CPI] applications.

The second generation of composite civil engineering structures will be developed using unique combinations of materials such as fiber/polymer composites in conjunction with concrete, steel, timber and other traditional materials of construction to create optimized hybrid structures for infrastructure applications. Long term, it is these combinations of materials, each contributing the most cost-effective menu of properties which can achieve wide-spread market acceptance. By building upon the long materials familiarity with and history of traditional materials and using composites to solve performance problems inherent in each of these materials (corrosion, rot, freeze/thaw, etc.) composites can make more rapid contribution to the civil engineering industry.

### The Importance of a Holistic View of Costs

The cost of composites compared to traditional materials is considered by many to be a critical issue! In fact, on the basis of cost per pound of load carrier, composite reinforcements are already lower cost than many traditional materials. Most of the time, composite products will displace or substitute for traditional materials. In the materials substitution process, the first step frequently requires developing an accurate definition of both the performance and cost targets for intended applications. For almost every application target, the existing specifications and standards for existing applications have been defined over time by the existing dominant material, as well as the all the cost "lore" associated with that particular end-use.

Normally the existing specifications/standards arise from the properties of the materials used rather than from the actual requirements of the application, (example: railroad ties tend to be defined by the properties of creosoted wood rather than by the structural requirements of loads and the operating environment).

It will be an important challange to take a close look at both performance and costs in order to establish realistic and meaningful criteria against which to evaluate potential composite applications. It may be necessary to develop new ways of measuring cost/performance such as comparing materials on the basis of normalized cost per unit of performance (pounds of load carried per dollar over the life-time of the installation, etc.) should help to clarify the long-term cost effectiveness of composites.

To look only at "first cost per pound" comparisons between materials is misleading and short-sighted. Many of the most important benefits of composites will come into play in indirect ways. Some of these indirect savings may include:

- jobsite labor savings
- reduced maintenance
- other jobsite advantages that translate into savings (damage, weight, etc.)
- life-cycle cost comparisons vs. first cost-only evaluation
- corollary savings
- parts consolidation
- pre-fabricated structures
- enabling technologies
- lighter weight structures, higher loads

For example, the cost of reinforced concrete along the mid-Atlantic coast is generally considered to be $800 per cubic yard. Of this figure, $700 per cubic yard represents the costs of labor & formwork compared to $50 per cubic yard for the concrete and reinforcing steel. According to Dr. Lawrence Bank of The Catholic University in Washington, D.C., an expert in this area, labor and formwork is generally considered to represent approximately 60-70% of a typical reinforced concrete job cost. Construction productivity is the key concept. It is certainly within the industry's technical capabilities to develop a form of specialized onsite fabrication of composite reinforcing rods produced in a mobile production facility mounted in a trailer with controlled environment. Such a product could be handled in labor-saving, high productivity methods on the jobsite to effect extraordinary potential savings and increased productivity.

Typically, when composites are substituted for traditional materials, the sequence of events is:

1. existing specifications are written around the materials properties of the current material. it is necessary to redefine the application on the basis of performance required (rather than a one-for-one substitution).

2. costs for the target application must be fully re-defined.

3. The specifying audience (architects, engineers, contractors, etc.) lack in-depth knowledge of the structural properties and design protocol for composites. It is unlikely that an optimized composite product for reinforcing concrete will resemble a

current steel reinforcing rods. Therefore, a program of practitioner education will be required.

4. Demonstration projects are one important way to promote industry knowledge and acceptance. However, it is critical that development initiatives must establish a balance between high-profile "Moon-Shot" breakthrough projects verses incremental demonstrations of more mundane applications (pilings, sheet pile, bridge decks, etc.). For every bridge across the Straits of Gibralter, there will be thousands of conventional bridges built, and tens of thousands of existing bridges rehabilitated or upgraded.

### Program Implementation Considerations

It will be important for composite development efforts to attack at the places where these versatile materials can do the most good. Initial targets which have already been identified include structural piling, sheet pile, dowel rods, reinforcing rods, pre and post tensioning tendons, cable stays, bridge decks and stay-in-place formwork.

Pressure-treated wood including, (creosote, CCA and CCR) have begun to pose serious long-term environmental concerns for in-situ contamination of water and soils as well as disposal. Environmental considerations associated with cutting down prime timber should become more than less compelling. The short service life of some traditional materials in more hostile environments (examples include treated wood pilings which have seen their service expectations reduced from 15 years to 7 years due to an unexpected resurgence of attack by marine wood borers) are excellent examples of overarching shifts which create new application opportunities for composites.

Composites have an inherent advantage over traditional materials in that composites are environmentally inhert. Composites do not rust, rot or suffer from the deterioration mechanisms that affect traditional materials. R&D should include developing programs to incorporate the maximum post-consumer and post-industrial recycled materials content as well as to engineer zero-release of toxic chemicals or byproducts of deterioration in composite products for civil engineering applications.

Readers should bear in mind that composites can be combined with other materials to enhance the properties of each. Examples might include development of composite structural marine pilings which use wood or concrete as structural cores.

### What Can Be Done With Existing Composite Material and Processing Technology

Existing composite materials and processing technology is capable of providing an extraordinary array of new applications for civil engineering structures. The new Civil Engineering Research Foundation (CERF) initiative to establish an Advanced Composites National Coordinating Council (NCC) is arguably the most important development in recent years. This initiative should concentrate on demonstrating and optimizing off-the-shelf technology as one of its highest priorities. For example, structural grating is a well-established product in the chemical processing industry, where it has been used for over thirty years in the most demanding installations. Yet, public sector engineers and maintenance officials seem unaware of these materials and their benefits. The potential exists to create near-term demand for off-the-shelf composite products while helping to create broad-based interest in these materials for new applications yet to be developed.

Existing technology should be capable of developing new applications in the following applications with minimum materials or processing extensions required:

1. Composite reinforcements for concrete

2. Composite structurals (beams, paneling, stay-in-place formwork, piling, sheet pile, etc.)

3. Non-structurals (re-useable formwork, facia, etc.)

At the same time, ASCE, the Composites Institute, ACI and others involved in the NCC initiative must begin to address strategic long-term issues such as new technology (materials and processes), knowledge building, or other critical issues including:

• Total Structural Knowledge (TSK) of specific materials systems and applications
• Structural Design Systems (SDS's) for each major material and application
• Engineering cirriculum development to teach students and practitioners about composites
• Specialized manufacturing processes
• Optimized combinations of materials (traditional composites as well as hybrids)
• Life-cycle cost techniques
• The Technology Transfer process itself including:
    - develolpment of stand alone parts/end-use product systems
    - performance requirements (identifying and quantifying same)
    - the fact that optimized composite design is almost always different than the design for the traditional material
    - knowledge exists, but it tends to be closely held and not in the form that civil engineers understand or feel comfortable with.
    - relationships between interested industry influence factors
    - focus and control of the development process itself vs. fragmentation and proliferation.

It is important to recognize that Technology Transfer is a process which encompasses a number of functions over time. One way to look at the technology transfer needs of the industry is shown as Figure 1 (below).

## Summary

In summary, the vision of the NCC initiative will be to select applications where there is the greatest need based on the limits of traditional materials in given operating environments, and where composites can bring the greatest value-added to the application (e.g., structural pilings & sheet pile, etc.). Fresh approaches will be required to create new design capabilities including:

- Design protocols for specialized applications
- Establish and promote new triad/quadratic engineering/design team comprising:
    - composites materials specialists      - civil engineers
    - designers                              - contractors
- Modular construction (most common elements, "Lego-type products", etc.)

The future ability of composite materials to help solve the nation's infrastructure and civil engineering problems will depend in large part on the success of the NCC initiative to establish cooperative programs between all parts of the civil engineering construction industry who stand to benefit from wide-spread application of composite technology. No single group (academia, practitioners, trade & professional organizations, government, etc.) can undertake and accomplish this job alone. Only by true inter-agency, inter-industry cooperation can the task be undertaken. But the results should be will worth the efforts in terms of new cost-efficient, low maintenance, long service life civil engineering structures... based on structural composites.

Figure 1

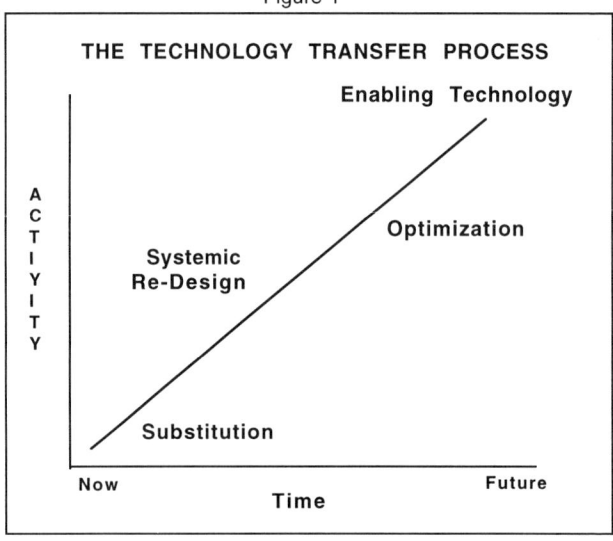

# COMPOSITE MATERIALS AS CONCRETE REINFORCEMENT IN THE NEXT DECADE

Salem S. Faza[1] and Hota V.S. GangaRao[2]

Abstract

Polymer matrix composites (PMCs) are researched for possible implementation in the construction industry. A high-volume market for the mass-produced, low-cost glass or carbon fiber reinforced polymer matrix composite reinforcing bars lies in the construction of highway and bridge structures, buildings, water and sewage treatment facilities, mass transit systems, and marine structures among many other constructed facilities. A number of factors affect the widespread use of FRP composites in the civil engineering construction. Recognizing that composite rebars in concrete cannot be designed as steel rebars, the Constructed Facilities Center (CFC) through an extensive research effort on concrete beams and decks reinforced with fiber reinforced plastic rebars (FRP), has developed design equations, charts, and tables for concrete reinforced with FRP rebars. This paper summarizes the recent developments in the use of fiber reinforced plastics bars as reinforcing elements in concrete.

Introduction

The backbone of America's commerce and industry consists of constructed facilities including public buildings, highways and bridges, air ports, transit systems, and many others. Most of our constructed facilities are deteriorating at a rate faster than our ability to renovate them. Recent studies project total needs of over $3 trillion during the next 15 years to bring these facilities to efficient operating levels. The declining productivity trends (3.3% productivity loss over four years according to the U.S Department of Transportation) can be reversed through the development and use of advanced composite materials in conjunction

---

[1] Research Assistant Professor, [2] Director, Professor, Constructed Facilities Center, West Virginia University, Morgantown, WV 26506-6101

with or in lieu of conventional materials, because conventional materials are known to have short life spans under harsh environmental conditions.

A composite material is composed of two or more constituents (e.g., Resins: polymers, ceramic, carbon; Fibers: glass, aramid, ceramic, carbon) coupled together to create an advanced material with engineering properties superior to those of its constituents. It is estimated that the composites industry will become a $90 billion industry by the year 2000 [1]. The composites industry will encompass an array of construction markets including highway transportation, underground storage tanks, off-shore structures, geosynthetics to control waste leaching, modular housing, and others. In addition, the composites industry will supply a variety of hybrid materials and structural shapes and systems that will be used extensively by the construction industry.

Objective

Advanced materials such as PMCs are rapidly invading a market once exclusively the domain of traditional materials such as metals. Advanced materials are being researched for possible field implementation and improvements in the longevity of facilities, and to build structures for noncorrosive and nonmagnetic environments. In this paper, the intrinsic issues that need to be accounted for in the design of concrete systems with FRP rebars are described.

FRP Bars for Concrete Reinforcement

The use of FRP ribbed or sand coated bars in lieu of mild or high strength steel as reinforcing or prestressing material in concrete is expected to extend the service life of constructed facilities because of better resistance to corrosion and higher damping properties than steel. In the United States, glass FRP bars, shown in Figure 1, are manufactured through a pultrusion process. The FRP bars are characterized, [2] and the results are summarized, herein. In addition, experimental results obtained by the CFC and the corresponding theoretical correlations are summarized in terms of elastic and ultimate bending moment, crack width, and bond strength of concrete beams reinforced with continuous glass FRP bars.

Mechanical Properties of Bars

Mechanical properties of FRP bars under tension, compression, torsion, and bending have been obtained by the CFC and others. The test specimens were made of thermoset materials with fiber volume fraction varying from 35 to 70%.

*Tension:* E-glass FRP bars under tension have a linear stress-strain relation up to about 96% of the ultimate strength with the average tensile stiffness of about $7.0 \times 10^6$ psi for 55% fiber volume fraction. The ultimate tensile strength of glass

FRP bars is dependent on fiber content, bar diameter, quality control in manufacturing, and type of matrix. Experimental results reveal that as the bar diameter increases, the ultimate tensile strength of the bar decreases, i.e., 134 ksi for 3/8" dia. bars to 88 ksi for 7/8" dia. bars. Such a decrease in strength is attributed to inadequacies in interlaminar shear transfer from outer fibers to the core fibers.

Figure 1  Sand Coated FRP Rebar for Concrete Reinforcement

*Compression:* The combination of matrix cracking and fiber kinking are observed for rebars loaded under compression. Unlike the tensile stiffness, the compressive stiffness varied with rebar size, type, quality control in manufacturing, and length to diameter ratio of the test specimen. The ultimate compressive strength values for 55% volume fraction of continuous E-glass fiber and vinylester resin is about 45 ksi. These test results are based on specimens that are under one inch in length, i.e., buckling effects are avoided.

*Bending:* Flexural strength and stiffness were obtained from three point bending tests. The dominant failure mode is observed from matrix cracking with no buckling of fibers in the compression zone. An average flexural stiffness for all bar diameters in tension is $6.8 \times 10^6$ psi, and the average flexural stiffness in compression is $6.0 \times 10^6$ psi. Ultimate flexural strength varied greatly depending on the specimen diameter, i.e., 131 ksi for 3/8" dia. bar to 61 ksi for 7/8" dia. bar.

## Theoretical Modeling of Bars

Theoretical modeling of continuous glass FRP bars to determine the mechanical properties under a variety of static loads has been attempted through micromechanical modeling, macromechanical modeling, and three-dimensional finite element modeling [3]. In the macromechanical modeling, FRP bars are treated as homogeneous but anisotropic straight rods of circular cross section. The theory of elasticity solution for circularly laminated rods was used by Wu (1992). Also, a three-dimensional finite element method, (FEM), using isoparametric elements and constitutive equations of monoclinic materials was employed by Wu. The ultimate strength predicted by the 3-D FEM is 25% higher than the experimental value and the theory of elasticity solution results in large discrepancies with the experimental values. To overcome the limitations of the above theories, GangaRao, et. al, [2] developed a mathematical model using the strength of materials approach including the shear lag between the fibers in the form of a parabolic strain distribution across the section. The model predicts tensile forces in the core fibers lower than those forces at the surface of the bars. A comparison of theoretical and experimental results under uniaxial tension and three point bending is found to be very good, i.e., experimental values are within 8% of our proposed theory.

## FRP Bar Performance as Concrete Reinforcement

*Ultimate Moment Capacity:* The phenomenon of developing moment resistance in FRP reinforced beams is identical to that of concrete beams reinforced with steel rods, provided that adequate bond between FRP and concrete is developed. The ultimate moment capacity of FRP reinforced concrete beams is given by Faza and GangaRao [2]. In order to take advantage of the high tensile strength of FRP bars, high strength concretes (6-10 ksi) should be used to maximize bending resistance of the structure. Bond, shear and compression failure must be avoided to attain the bending mode failure. Technical information published by GangaRao and Faza [4] should be reviewed to get a more comprehensive understanding on bond strength of FRP bars embedded in concrete.

*Crack Width Analysis:* A number of equations have been proposed [5] to predict crack widths in flexural members with steel rebars. The ACI 224 Committee Report on crack control of steel reinforced concrete beams has suggested the following parameters that have significant influence: 1) stress in the reinforcement; 2) concrete cover thickness; 3) area of concrete surrounding a rebar; and 4) strain gradient from the rebar to the tension face.

The current ACI 224 equation for crack width cannot be used to predict crack width in concrete beams reinforced with FRP bars. The effects of bond strength between concrete and FRP bar and the low FRP bar modulus of elasticity need to be accounted for while establishing crack widths. The strains in FRP bars are expected to be four times those of steel because steel bars are four times stiffer than FRP bars. Since crack spacing is governed by the bond stress between the FRP bar and concrete, higher bond strength of sand coated FRP bars [4] reduces crack spacing to half the crack spacing of a steel reinforced beam.

To achieve realistic crack width calculation, a crack width equation that incorporates the actual bond strength of FRP bars is needed. Such an equation must be obtained from the experimental data until the FRP bar manufacturing process is standardized. An expression for crack width for FRP reinforced concrete beams has been derived by multiplying an average strain in FRP bars with the expected crack spacing, where the crack spacing is a function of bond stress, bar diameter, and tensile strength of concrete. It should be noted that the low modulus of sand coated FRP bars together with higher bond strength will produce crack widths that are about two times larger than in steel reinforced concrete beams. Therefore, the use of FRP rebars as concrete reinforcement require a review of the existing crack control criteria.

Figure 2  Bending Test Set-up of Concrete Beam Reinforced with FRP rebars

*Deflection Analysis:* Based on the mechanical properties of FRP bars, the response of concrete beams reinforced with FRP bars was investigated in terms of pre- and post-cracking load-deflection behavior. The use of sand coated FRP bars in addition to high strength concretes improved the ultimate moment capacity, and crack width and propagation of concrete beams.

Theoretical correlations with experimental deflections were found to be excellent when they were based on a modified moment of inertia in the post-cracking zones due to the lower modulus of elasticity of the rebar. The modified moment of inertia that is proposed by Faza and GangaRao [4] incorporates both the cracked moment of inertia as well as the current ACI method to compute post-cracking moment of inertia of concrete beams. In addition, the modified moment of inertia accounts for the crack pattern and extent of cracking in a concrete beam reinforced with FRP rebars

### Future of Composites in Concrete Reinforcement

Composites have significant role to play for constructing modern infrastructure systems. They also play an important role in rehabilitating existing systems, such as bridge decks. It is now well recognized that traditional materials such as steel and concrete, in their present form, cannot be made to perform satisfactorily due to their susceptibility to environmental effects and consequent degradation. Thus, there is a place for composites as a new and emerging technology for solutions. In essence, use of composites in construction is an innovative technology. Thus challenging projects, such as long span suspension bridges can be designed and built in the future.

Based on the comprehensive study conducted at the Constructed Facilities Center on the future of composites in construction, the following highlights are provided:

1. Non-corrosive, high strength composite FRP rebars represent one of the advanced materials of the future. However, the design and performance requirements of these advanced materials are very stringent. The real challenge is to make sure that they are available in high volume and at competitive costs in comparison with conventional materials. As in the case of any new technology, there is a need for improved understanding among engineers and contractors of the behavior of these advanced materials.

2. For optimized use of polymer composites in construction, it is necessary to consider innovative material processing and construction technologies by appropriate changes in building codes, standards, and design practices. Wherein, industry, government and university personnel have to function as a team to promote the use of composites in construction.

3. A systematic strategy is necessary for enhancing the knowledge-base of composites and their overall implementation in research and development, manufacturing, design / construction / inspection, environmental considerations and technology transfer.

4. To satisfy performance requirements, it is necessary to understand material parameters, design aspects at component as well as system level, full-scale structure response and the life-cycle costs.

The above information suggests that there is an immense potential for use of composites in the construction of diversified structural systems.

Summary

Use of fiber reinforced plastics as reinforcing elements in construction has been limited because of several factors. In general, lack of design guidelines, material properties, standardized test procedures, and design awareness by the construction/structural engineers. Attitudes are changing rapidly, and substantial growth in the use of FRP in construction is observed.

Reinforcement for concrete with FRP rebars has been proved to be feasible and economical. In the past applications were limited because of the poor quality of rebars, smooth surface of rebars and associated poor bond strength with concrete, and lack of design guidelines other than those of the ACI for steel reinforcement.

However, salient characteristics of FRP rebars, mainly their noncorrosive and nonmagnetic nature, motivated practicing engineers to use FRP rebars in advanced reinforced concrete structures. The dramatic deterioration of U.S. constructed facilities as a result of corrosion, motivated further research in this area. As a consequence, substantial improvements in the product have been achieved. Particularly important are the utilization of ribbed, sand coated rebars with high strength concrete and the development of design guidelines that depart from direct replacement of steel by FRP rebars.

It is recognized that composite structures cannot be designed as steel or concrete structures. The successful growth of FRPs depends on proper accounting of their material properties and innovations in designs. Advances have been made in recent years in terms of FRP applications for infrastructure construction. However, additional developments of fiber composites are needed to continue to improve the design tools, materials and products to enhance the efficiency of composites in construction for the 21st century.

## References

1. FMS (Federation of Materials Societies), "Creating a Materials Future: Translating Scientific Leadership into Commercial Advantage," Washington, D.C., 20036, 1989.
2. Faza, S.S., GangaRao, H.V.S., "Glass FRP Reinforcing Bars for Concrete," Chapter in *FRP Reinforcement for Concrete Structures: Properties and Applications*, Nanni, Editor, Elsevier Science Publications, New York, NY, 1993.
3. Wu, W.P., "Thermomechanical Properties of FRP Bars," *Ph.D. Dissertation*, West Virginia University, 1991.
4. GangaRao, H.V.S. and Faza, S.S., "Bending and Bond Behavior and Design of Concrete Beams Reinforced with FRP Rebars," *WVDOH-RP-83 Phase I Report*, 1992.
5. Halvorsen, A., "Code Requirements for Crack Control," *Proceedings of the ACI Fall Convention*, Lewis H. Tuthill International Symposium on Concrete and Concrete Construction, 1987.
6. GangaRao, H.V.S. and Barbero, E., "Construction Applications," Vol. 6, *International Encyclopedia of Composites*, VCH Publishers, Editor S.M. Lee, New York, NY, 1992.

# PULTRUDED COMPOSITES: *MATERIALS FOR THE 21ST CENTURY*[1]

**Ayman S. Mosallam**[2]**, Ph.D., P.E.**
*The Composite Structures Group*
*Civil, Mechanical, and Environmental Engineering Department*
*THE GEORGE WASHINGTON UNIVERSITY*
*WASHINGTON, DC 20052*

### ABSTRACT:

*Pultruded fiber reinforced plastics (PFRP) composites are considered to be one of the high-performance construction materials (CERF (1993)). These materials and systems have been recognized as the potential materials for repairing and constructing our infrastructures for the 21st century. The main objective of this paper is to introduce pultruded composites to the civil engineers and to create the awareness of its important applications. A description of the pultrusion process, material composition, and the different mechanical properties of PFRP composites are presented. Some of the civil engineering applications where pultruded composites are considered superior systems are described. The paper describes also the structural behavior of commercially produced pultruded shapes which includes both the weakness and the recommendations for increasing the structural capacity of PFRP open-web profiles. A review*

---

[1]For presentation at the ASCE Dallas Convention "*Moving into the 21 Century*", October 25-28, 1993.

[2]On sabbatical leave from the Faculty of Engineering, Shoubra, Zagazig University, Egypt.

of the research work on the performance of PFRP frame connections, creep behavior, and the dynamic response of these materials is presented.

## INTRODUCTION:

For the past few decades, aerospace industry has been a major user for advanced composites materials for structural applications. Recently, civil engineers and the construction industry begin to realize the potential of these materials in providing remedies for many problems associated with the deterioration and corrosion of infrastructures. One of the popular form of structural composite materials for civil engineering applications is known as "FRP" or fiber reinforced plastics. These materials are composed of high performance fibers (glass, carbon, aramid,..) alone or hybrid, embedded in polymer matrix (e.g. polyester, vinylester, epoxy,..). These materials are being produced commercially in a number of standard "off-the-shelf" structural shapes (FIG. (1)). These shapes are similar to those being produced by the steel industry and are currently manufactured by several manufacturers in the United States, Japan, Canada, and Europe.

This paper serves as a state-of-the-art report on the applications and developments of FRP structural shapes for civil engineers. The paper describes the different available structural shapes which includes: thin-walled open and closed web profiles (e.g. I, H, C, Box,..), FRP plates, gratings, and grids. A description of some existing civil engineering projects which were built entirely from these structural shapes are also be introduced. In addition, a review of research work in the area of material mechanical characterization, connections, and full-scale testing of FRP frame structures is presented. Recommendations and suggested details are provided for improving the existing pultruded fiber reinforced plastic structural shapes and consequently to increase their structural capacity, reliability, and efficiency.

## WHY PULTRUSION?

The pultrusion process is a proven manufacturing method for obtaining high quality FRP parts having consistently repeatable mechanical properties (

Werner (1984)). A number of recent articles has discussed various aspects of the pultrusion industry and the use of pultruded parts for variety of structural applications (Mosallam (1990), Sims et al (1987), and Sumerak (1985), . Some of the basic features of the pultrusion process are; complex shape and length capabilities, precise positioning of reinforcements, low labor and tooling cost, low scrap factor, and variable wall thickness capabilities. The raw materials which are used in the pultrusion process are a liquid resin mixture which contains the base resin, catalyst, pigment, fillers, and other desired additives and flexible reinforcing fibers (e.g. E-Glass, Carbon,..). In the pultrusion process, reinforcing fibers in the form of rovings or mat are pulled through a liquid plastic resin. Next, these impregnated fibers are pulled through curing and forming dies and exit the process as finished fiber reinforced plastic structural sections. These sections can be cut to any desired length. The volume fraction and fiber orientation, resin system can be tailored to meet special design requirements. FIG. (2) describes the different steps of the pultrusion process.

**ADVANTAGES:**

Some of the attractive and unique features of PFRP structural composites are their durability and resistance to the marine environment; their toughness, particularly at low temperatures, their vibration damping capabilities, their energy absorption under earthquake loading, their electromagnetic transparency, their low value of coefficient of thermal expansion in some cases, pigmentability and decorative characteristics, and their high strength-to-weight ratio. It is expected that these unique properties can be used to produce an optimum structural system with minimum *life cycle cost*.

Some of the civil engineering applications where pultruded composites are considered to offer advantages over other conventional construction materials, can be classified into the following:

1) **Aggressive Environments**

 - Water and waste-water treatment plants structural and non-structural elements (e.g. weirs, sludge flights, ..),

- Water desalination plants structures
- Water front structures
- Offshore structures (Off shore oil rigs, marine risers,..)
- Cooling towers (housing units)
- Petrochemical and nuclear power plants
- Paper and pulp mills
- Chimneys
- Agricultural and irrigation equipment and structures (e.g Manure filters, Weirs, barn structures, water gate guides,...)

2) **Enhancement and Retrofit of Existing Infrastructures and Historical Buildings:**

- As reinforcement of existing structures, a lower dead load is expected due to the higher specific strength and stiffness of special types of pultruded composites. This will result in an increase in the live load capacity of the structure.

- *Pre-earthquake* strengthening elements for structures which were not designed to withstand the current local local seismic loads. For example, R/C bridge columns, girders, slabs and other infrastructures. It is also possible to increase both the ductility and strength of the existing structure with little additional weight of PFRP composites.

- *Post-earthquake* rehabilitation materials for reinforcing and strengthening damaged R/C, wood, and masonry structures, and for upgrading the capacity of historical buildings and structures. Thin composite sheets or plates made of FRP composites increase ductility, stiffness, strength and, consequently, an increased capacity to withstand future earthquakes.

**Electromagnetic Transparent and Low Conductive Structural Systems:**

- Highway auxiliary structures (Light poles, guard rails, signing structures,..etc.

- Special hospitals structures
- CB Radio Antennas
- Radar and military facilities,
- Airport control towers,
- Electrical and telecommunication manholes,
- Electrical cable trays
- Transformer spacing sticks
- Third rail cover boards
- Transmission and electrical power towers, and power plants structures.

The latter application is recently under investigation to verify the potential of FRP composites *environmentally* when utilized as the main structural system in the electrical tower application. The non conductive PFRP is expected to reduce the health hazards associated with the presence of electric power in *steel* towers in urban areas.

In the above examples, pultruded composites offer the advantages of corrosion resistance and can be included as a complete *system* or *subsystem*.

**Structural Behavior of Pultruded Shapes:**

As it was discussed earlier, the pultrusion process is somewhat unique in that it allows for combining a variety of reinforcement types in the same section (Martin and Sumerak (1978)). Most of the commercially produced PFRP structural shapes are composed of multilayers of surfacing veil or Nexus, continuous fibers (roving), and continuous strand mat. The typical volume fraction of fibers for "off-the-shelf" sections is in the range of 40% to 45%. A variety of structural profiles (open and closed-web) are now available similar to steel sections (H, I, C, L, ...). The major reinforcements of these sections are concentrated in the longitudinal direction of the section with minimum reinforcements in the transverse direction.

For the past five years, a number of full-scale experimental studies on the mechanical characterization of PFRP structures has been performed (Mosallam (1990), Barbero (1991), Daniali (1990), and others). The results obtained from these studies demonstrate the necessity for optimizing the PFRP structural shapes. For example, studies conducted by Mosallam (1990) indicated that a premature local failure of *unstiffened* PFRP open-web sections occurs well before the ultimate bending capacity of the beam cross section is reached (see FIG.(3)). This common local failure is due to both the low level of reinforcement at the web/flange intersection and the discontinuity of these fibers near the center line of the section. FIG. (4) shows the *inverted triangular* area which contains minimum or no reinforcement of some of the commercially produced open-web sections. A recent study conducted by the author, indicated that this reinforcement discontinuity is responsible for approximately 50% reduction in the flexural capacity based on material strength of the cross section. According to this study, an inefficient and uneconomical design of the composite section will result if no stiffeners are included in the design to increase both the buckling and postbuckling behavior of the thin-walled sections. This inefficient design approach leads to cost-ineffective members. In addition, this premature failure caused by the separation of the web and the flanges of the open-web PFRP elements at locations with high stress concentration (usually at the connections and the girder mid-span) will affect the general behavior of structure. For example, results of experimental and theoretical research work (Mosallam (1993), Mosallam, Abdelhamid, and Conway (1993)) showed the direct effect of this premature failure of the column open-web section in the significant loss of the rotational stiffness of PFRP frame connections as the crack grows up to the failure. The loss of the connection flexural stiffness is responsible for the continuous increase in the positive flexural stresses at the girder mid-span. Consequently, a reduction in the critical buckling stresses of the thin-walled PFRP sections will be expected(Mosallam (1990)).

A discussion on the available techniques using PFRP *transfer elements or continuous threaded rod* systems to improve the structural efficiency of existing PFRP sections is discussed in detail by Mosallam (1993). FIG. (5) shows some of these strengthening details for existing PFRP structures.

*PFRP Frame Connections:*

For any frame structure, connections are considered to be critical structural elements because they play a major role in controlling both the serviceability and the ultimate strength of structure. Careful design of the connecting elements will ensure both the safety and the efficient use of the material.

Previous studies on PFRP frame structures (Mosallam (1990) & Bank, Mosallam, and Gonsior (1990)), showed that *a premature failure of pultruded shapes will occur if a wrong connection detail is used* (FIG. 6). Based on this, Bank, Mosallam and McCoy (1992) have extended this work by introducing different connection details that are designed overcome the premature failure of the pultruded shapes at the web/flange junction of PFRP H-beams. The connection details presented in their study considered the anisotropic properties of the PFRP structures. Their results showed that maximum strength and maximum stiffness can be achieved by using connection details where both mechanical and adhesive elements are used.

All PFRP connections details which were developed and tested in previous studies have utilized PFRP connecting elements which are commercially produced and were not intended specifically as connection elements. This was an appropriate approach to investigate the efficiency of the existing connection details, as well as to provide strengthening details for reinforcing the existing PFRP connections. However, improved approaches for connecting PFRP structural elements with maximum efficiency and safety is needed. The design criteria for these connecting elements include, proper fiber orientation, ease of erection and duplication, geometrical flexibility to be adapted to the majority of structural sections and connections, and to maximize both the overall stiffness and ultimate capacity of PFRP frame connections. Based on these criteria, a FRP prototype connector was developed (FIG.(7)) and fabricated (using resin transfer molding, RTM) from a E-glass/vinylester composition. This FRP connecting element (designated, herein, as the Universal Connector " UC") was developed by Mosallam (1993), The UC element, can be used for the majority of PFRP connection details for joining different structural shapes, e.g., exterior and interior beam-to-column connection (FIG. 8), column-base connections, continuous beam connections, beam-to-girder connections, and others. An extensive theoretical and experimental program in the area of PFRP connection development and characterization is in progress (Mosallam, Bedewi, and Goldstein (1993)).

*Viscoelastic Behavior of PFRP Structures:*

Unlike steel, all PFRP composites are viscoelastic and, as such, display time-dependent stress-strain behavior, as do concrete and wood. Viscoelastic materials exhibit elastic action when loads are applied followed by a slow and continuous viscous increase of strain at a decreasing rate over time. Upon removal of load, these materials exhibit an initial elastic recovery followed by a similar continuous decrease in strain. Commercially produced PFRP composites display significant viscoelastic effects on high temperature exposure, especially when loaded continuously. That is, as the temperature increases, FRP composites will experience losses in stiffness (and also strength). For applications where high temperature environment is expected, special resins, admixtures, and reinforcements should be selected.

The viscoelastic behavior of pultruded FRP materials is greatly affected by the creep characteristics of the matrix material which distribute the stresses among the individual fibers and controls the interlaminar shear and extension perpendicular to the fiber direction (Mosallam and Bank (1990)). This viscoelastic behavior is also dependent on the creep behavior of the reinforcing materials which serves as the primary load-carrying constituent. However, these effects are negligible for glass and carbon fibers. A summary of the research work in this area is presented below.

Holmes and Rahman (1980) performed a creep study on hand-made rectangular simply supported beams bidirectionally reinforced with glass fibers. In their work, the beams were subjected to various loading procedures and resulting creep parameters were presented. Brinson *et al.* (1980); and Hiel and Brinson (1983) and Dillard and Brinson (1983) presented numerical procedures for predicting creep and delayed failures in laminated composites. In their work, a time-temperature-stress-superposition (TTSSP) was employed. This superposition was based on the fact that certain environmental conditions such as temperature and stress serve to accelerate the deformation process associated with the viscoelastic properties of composites. Similar work was performed by Yen and Williamson (1990) on the accelerated characterization of creep response of composite materials. The effect of mechanical prestrain on the creep response of hybrid composite at elevated temperature was studied by Williams and Pindera (1991). Beckwith (1984) performed an experimental and theoretical study on the creep behavior of Kevlar/Epoxy composites. In this work, creep deflections (and not the creep strains) were represented by a power law and it was concluded that the creep behavior in the laminate composites is primarily "fiber-dominated" and independent of resin modulus for the same loading condition. Hollaway and Howard (1985) investigated the behavior of a double layer skeletal structure made from pultruded composites. For pultruded composites, Daniali (1989, 1990) performed an experimental study on the time-dependent behavior of two types of FRP lintels at ambient and elevated temperatures. Recently, Levy and Murray (1993) conducted a creep study in several coupons of glass-reinforced polyester and vinylester which were cut from both webs and flanges of commercially produced PFRP composites. In their study, less creep for vinylester specimen as compared to polyester specimens were reported.

A comprehensive study on the creep behavior of PFRP composites was conducted by Mosallam (1990). Both experimental and theoretical investigations on the time-dependent response of a PFRP portal frame were conducted (see FIG. (9)). Three creep tests were also performed on companion coupons (FIG. (10)) and the results were compared to the results obtained from a PFRP full-sized creep test. Using the creep strain data along with Findley's creep model, two viscoelastic moduli for pultruded FRP material were obtained. The viscoelastic moduli are expressions which describe the stress-strain relationships of viscoelastic materials. They are the initial slope of the isochronous stress-strain curves, as obtained under sustained stress or strain (SPDM (1984)). For design proposes, simple expressions for the viscoelastic moduli are needed for long-term structural analysis of FRP structures. For this reason, a linear viscoelastic behavior of the material is usually assumed, i.e. the stress is proportional to strain at a given time. This assumption of linear viscoelastic behavior for plastics is valid only under service loading where the stresses are relatively low. On the other hand, high stress levels, elevated temperature and severe environmental conditions increase the non-linear behavior of the plastics (SPDM (1984)). In these situations, non-linear viscoelastic moduli must be used which are dependent on both time and stress level. In this case, analysis of a simple structure will require an infinite number of viscoelastic moduli since the stresses will vary along both the cross section, as well as along the length of the structure. However in practice, this approach is seldom necessary or justified.

The following conclusions were reached for the specific PFRP sections examined based on the experimental and the analytical results; 1) a significant portion of the creep occurs during the first 2,000 hours of loading and creep parameters obtained from this part can be used to predict the long-term behavior of the FRP frame; 2) Creep parameters obtained using coupon creep tests can be used to predict the viscoelastic behavior of FRP structures; 3) The rate of creep for shear strain is larger than the rate of creep of the tensile strains for unidirectional PFRP sections; 4) The compression strain creep behavior is different and lower than tensile creep behavior for unidirectional PFRP sections; 5) Findley's power law can be used successfully to describe the creep behavior of pultruded FRP structures; 6) the simplified assumption of linear viscoelasticity can be used to obtain simple expressions for the viscoelastic moduli of FRP materials; 7) The inclusion of the FRP beam-to-column flexibility, as well as, shear deformation effects is crucial in predicting the long-term response of FRP frame structures; 8) The principle of superposition can be used to describe the creep recovery of FRP structures. A detailed discussion on the creep behavior of PFRP structures is given by Mosallam (1990) and the ASCE Structural Plastic Design Manual (1984).

*Dynamic Behavior of Pultruded Structures:*

The dynamic response of both PFRP materials and structures was investigated by Mosallam and Abdelhamid (1993). In this study, results of experimental dynamic tests of FRP pultruded structural elements and framed structures were presented. The thin-walled elements used in this study were standard "off-the-shelf" pultruded 4" x 4" x 1/4" H-beam and 2" X 2" X 1/4" square tube made of E-glass/polyester composition. All the connectors and connection elements were made of PFRP threaded rods, nuts, and high-strength epoxy adhesive. The test specimens in this study were excited dynamically using both impact loading and shaking loads. Experimental modal analysis was used to extract the natural frequencies, modal damping, and mode shapes of the test specimens. Comparison between two types of frame connections were also performed to determined the effect of using high-strength adhesives. The study further showed the validity of using both the material properties and the lay-up of the coupons in modeling PFRP beams and frame structures. FIG. (11) shows the test setup of the PFRP dynamic test. Based on this study, further studies on the acoustical properties of the PFRP sections were recommended.

## EXISTING STRUCTURAL APPLICATIONS:

In the past ten years or so, an increase in the level of awareness of the pultruded composites and their potential in the civil engineering applications is apparent. Several projects have been constructed entirely using pultruded fiber reinforced sections as the main structural system. One well publicized application is the four PFRP turret towers on top of the *Sun Bank* Building, Orlando, Florida. The base dimension of each tower is 35' X 35' (10.67 m X 10.67 m) and its height was selected to house a 20' high (6.10 m) antennae for police and fire communications. FIG. (12) shows the three-story high tower framing which was built entirely from pultruded fiber reinforced plastics (PFRP) shapes (H, angles, threaded rods and nuts). All columns and girders were constructed using open-web H sections which were connected using FRP bolts and nuts. The connections are similar to those for steel framing, of the type discussed earlier. In this project, PFRP composites were selected because pultruded shapes provided electromagnetic transparency and low reflection level of radio-wave. In addition, this system was chosen for its favorable architectural quality.

FIGs. (13) show a complete platform structure constructed entirely from PFRP and grating panels. The system was chosen because of the superiority of FRP composite in harsh environments.

Due to the non-magnetic properties of PFRP composites, these materials are commonly used for facilities with sensitive electronic instrumentations. FIGs. (14) and (15) show complete frame structure which were constructed using PFRP materials. The ease of fabrication, transportation and erection resulted in shorter construction time and consequently a relatively lower *life cycle cost*.

One recently approved bridge project where PFRP composites represent the major part of the structural system is the Thames Foot Bridge, United Kingdom. The location of the composite bridge is between the Cotswold town of Lechlade and the Village of Buscot. The project is being funded by the government, private sector, and by fund-raising and sponsorship and will constructed in 1994 (Greenberg (1993)). The span of the bridge is 22 meters ($\cong$ 72 feet). The main structural elements of the bridge are made of light weight composites and diamond-coated film glass and carbon PFRP.

## BARRIERS TO ACCEPTANCE:

One of the major questions which needs to be answered for the structural engineer who has been dealing, for decades, with conventional materials such as steel, concrete is *why should s/he select this new material?*. The same question was asked forty years ago when composites were introduced to the aerospace industry. The answer to this question, especially to the civil engineering community, is not as simple as many engineers think. The resistance to acceptance lays on the *apparent burden* which this engineer expects to encounter and the sacrifices involved in dealing with more advanced design procedures and products. Some of these concerns are: i) the absence of authoritative codes and material specifications (Chambers (1991, 1992)) ii) the lack design procedures which, have been established for decades for conventional materials (concrete, steel, wood) ii) the direct involvement of the structural engineer in the manufacturing process, material *tailoring* and selection, and lastly iii) the need of relatively skilled labor at both the fabrication and construction sites as compared with conventional construction materials.

The effective tool to overcome the above roadblocks, in the author's opinion, is *education.* First, educating the construction industry about the nature of these materials and the associated benefits, as well as identifying their special mechanical properties such as the anisotropic and viscoelastic natures of PFRP composites. Second, establishing a modified or new civil engineering curriculum which includes composite design courses similar to those for steel, timber, and concrete at both the undergraduate and graduate levels. This step is essential in order to prepare a new generation of structural engineers equipped with the skills required for designing with advanced composites. This, off course, requires the establishment of a design code and specifications. Recently, positive movements by the different engineering and industrial organizations have been initiated to pursue this important task. For example, a new subcommittee of the ASCE Structural Composites and Plastics committee has been formed, in August 1992, to initiate an initial proposal for establishing an ASCE code for composites. The Pultrusion Industry Council (PIC) of the Society of Plastic Industry (SPI) has already has initiated the first phase of a structural design manual project. A detailed discussion on this subject is presented by Chambers (1993).

It is also important to identify several facts which will help in convincing the civil engineer who will be introduced to FRP composites for the first time. First, the fact that she or he for decades has been using some forms of composites with no difficulties (e.g. reinforced concrete, natural, and laminated wood). Secondly, the engineer should understand that the aim of using FRP composites is neither to replace nor to compete with other construction industries such as concrete, steel, wood,..etc. This misconception has led to both confusion and the creation of the defensive position by loyal concrete, steel, and wood users. The matter of fact is; composites are the right choice in some applications when other construction materials are disqualified in part or on whole. For this reason, the structural engineer should begin to look at different materials as elements of a complete "Structural System". In contrast, he or she should understand that these materials are here to enhance and assist the conventional materials in certain applications beyond their capabilities. In addition, these materials if used properly with other conventional materials can produce an optimum engineered structural "system" capable of solving many problems associated with our infrastructures. For example, fiber reinforced plates and sheets have been utilized for both strengthening and enhancing the structural performance of reinforced concrete, wood, and masonry structures Meier and Kaiser (1991), Saadatmanesh and Ehsani (1990).

**Conclusions and Recommendations:**

From the above discussion, a new strategy to deal with advanced construction composites is shown to be an important task. It is important those engineers who are willing to take advantages of these materials to learn the basic mechanics of composites. The directionlly dependent, the viscoelasticity, and environmental properties need to be clearly understood. The pultrusion industry, in particular, and the composite industry in general must support and the encourage academic and research studies. The great attention which was shown lately by the different professional and federal organizations in promoting and supporting research and demonstration projects are the first positive sign of accepting these new materials. For example, the recent activities conducted by the Civil Engineering Research Foundation (CERF), the Pultrusion Industry Council (PIC) of the Society of Plastic Industry (SPI), the US corps of engineers CPAR program, the Federal Highway Administration (FHWA), and the National Science Foundation (NSF) indicate the interest and the high level of awareness by the advanced engineering communities of the role that composites may ply a major role in improving the quality of our infrastructures. Design codes and material specifications are required in order for this material to enter and to survive the demand imposed by the construction industry. A vast quantity of research data has been generated in this area which needs to be collected in a *national data base system*. It will be importance to establish this system to evaluate the different research and demonstration works by forming a specialized technical committee to evaluate and select useful information and design data and use them in establishing unified design procedures.

We must continue to optimize PFRP for cost and performance effectiveness. One way of achieving the maximum efficiency is the use of stiffeners when designing PFRP structures. For example, shear stiffeners, flexural stiffening plates, bearing stiffeners, etc.. Although PFRP materials are different than steel and concrete in their mechanical properties, the objective of creating a safe, fit, and economical engineered system still hold. For this reason, the inclusion of such stiffeners in the complete design will able PFRP structures to compete with other materials.

Experimental and analytical research work on the long-term performance of PFRP under different environmental conditions is needed in order to increase the confidence level of these materials. Education at the university level and the practical training programs for structural engineers are essential.

As we enter the 21st century with this advanced materials, we need advanced design tools to achieve the optimum design and to ensure the required performance of PFRP composites. For other conventional construction materials, the use of simplified design formulas, charts, and tables were acceptable. Fortunately, and due to revolutionary advancement in personal computers technology and the availability of powerful computational techniques such as finite element method (FEM), the use of computers in designing composite structures will ensure the efficient utilization of composite sections. This is due to the fact that the optimum design and performance prediction of advanced composites is relatively complex and can only be achieved efficiently through more detailed calculations. Simplified equations are still important for quick estimate and as a tool for comparison. In short, advanced composites provides the structural engineers, for the first time, with the challenging opportunity to create innovative structural systems.

**Appendix I:** *(SI- Units)*

1 lb = 4.448 N  
1 inch = .0254 m

1 psi = 6.895 kPa  
1 ft = 0.3048 m

**References:**

Bank, L.C., Mosallam, A.S., and McCoy, G.T. (1992) " Design and Performance of Connections for Pultruded Frame Structures," Proceedings, The 47th Annual Conference, Composite Institute, The Society of Plastics Industry, Inc. February 3-6, Paper No. 2-B.

Bank, L.C., Mosallam, A.S., and McCoy, G.T. (1992) " *Make Connections Part of Pultruded Frame Design*", Modern Plastics Magazine, Ed. P. A. Toensmeier, McGrew-Hill, August, pp 65-67. (Also appeared in *Modern Plastics International*, August, pp. 34-36.)

Bank, L.C., Mosallam, A.S., and Gonsior, (1990). "Beam-to-Column Connection for Pultruded FRP Structures." Proceedings, ASCE First Materials Engineering Congress, Denver, Colorado, August 13-15, 804-813.

Beckwith, S.W. (1984). "Creep Behavior in Kevlar/Epoxy Composites." Proceedings, 29th National SAMPE Symposium, April 3-5, 578- 591.

Barbero, E., (1991) "Pultruded Structural Shapes: Stress Analysis and Failure Prediction" *in Advanced Composites Materials in Civil Engineering Structures*, Proceedings of the ASCE Specialty Conference, Las Vegas, Nv, Jan 31-Feb 1, S.L. Iyer and R. Sen. Eds., NY, NY, ASCE, 224-232.

Brinson, H.F., Griffith, W.I., and Morris, D.H. (1980). " Creep Rupture of Polymer-matrix Composites." The 4th SESA International Congress on Experimental Mechanics, Boston, MA, May 25-30, 329-335.

Chambers, R. E., (1991) "Composites in Construction Require a Structural Design System" Workshop on Advanced Composites for Offshore Structures, MIT, Cambridge, October, 30-31.

Chambers, R. E., (1992) "Composites in Construction Require a Structural Design System," Session on Composite Opportunities in the Marine/Waterproofing/Off-shore Industry, SPI Composite Institute's 47th Annual Conference and Expo 92, Cincinnati, OH, 3-6 Feb.

Daniali, S. (1990). "Time-Dependent Behavior of FRP Lintels." Proceedings, ASCE First Materials Engineering Congress, Denver, Colorado, August 13-15, 814-823.

Findley, W.N.(1987). " 26-Year Creep and Recovery of Poly(Vinyl Chloride) and Polyethylene," Polymer Engineering and Science, 27(8), April, 582-585.

Greenberg, S. (1993) "Spanning the Thames." The Architects Journal, EMPA, Busner Publications, UK, March, 19- 20.

Hiel, C.C., Brinson, H.F. (1983). " The Nonlinear Viscoelastic Response of Resin Matrix Composites." Composite Structures 2, Proceedings, The 2nd International Conference on Composite Structures, Paisley, Scotland, 14-16 September, 271-281.

"High-Performance Construction Materials and Systems: *An Essential Program for America and Its Infrastructures"* Civil Engineering Research Foundation (CERF) Executive Report, Report # 93-5011.E, April.

Hollaway, L., and Howard, C. (1985). "Some Short and Long Term Loading Characteristics of a Double Layer Skeletal Structure Manufactured from Pultruded Composites." Composite Structures 3, Proceedings of the Third International Conference on Composite Structures, Paisley, Scotland, 9-11 September, 788-808.

Hollaway, L., and Farhat, A. M., (1990) "Viberational Analysis of a Double-layer Composite Material Structure." Composite Structures, Vol. 16 (4), 283-304.

Holmes, M., and Rahman, T.A. (1980). "Creep Behavior of Glass Reinforced Plastic Box Beams." Composites, April, 79-85.

Levy, A. and Murray, P., (1993) "Creep Response of Pultruded FRP Materials", Proceedings, The ANTEC Conference,(Student Papers Session), Society of Plastic Engineers, New Orleans, Louisiana, May 9-13.

Martin, J., and Sumerak, J. (1978) "A Review of the Market for Pultruded Applications-And Factors Affecting its Growth" Proceedings, the 38th Annual Conference, Reinforced Plastics/ Composite Institute, The Society of the Plastic Industry, Inc., February 7-11, Session 6-F, pp 1-5.

Meier, U., and Kaiser, H., (1991) "Strengthening of Structures with CFRP Laminates" *in Advanced Composites Materials in Civil Engineering Structures*, Proceedings of the ASCE Specialty Conference, Las Vegas, Nv, Jan 31-Feb 1, S.L. Iyer and R. Sen. Eds., NY, NY, ASCE, 224-232.

Mosallam, A. S., (1990) "Short and Long-Term Behavior of Pultruded Fiber-Reinforced Plastic Frame," Ph.D., thesis, Catholic University of America, Washington, D.C.

Mosallam, A.S., (1993) "Stiffness and Strength Characteristics of PFRP UC/Beam-to-Column Connections." Composite Material Technology, PD-Vol. 53, Proceedings, ASME Energy-Sources Technology Conference and Expo, Texas, January, 275-283.

Mosallam, A.S. and Abdelhamid, M.K. (1993) "Dynamic Behavior of Pultruded PFRP Structural Sections" Composite Material Technology, PD-Vol. 53, Proceedings, ASME Energy-Sources Technology Conference and Expo, Texas, January.

Mosallam, A.S. and Abdelhamid, M.K., and Conway, J.H. (1993) "Performance of PFRP Frame Connections Under Static and Dynamic Loads" Proceedings, the 48th Annual Conference, Composite Institute, The Society of Plastic Industry, Cincinnati, Ohio, February.

Mosallam, A.S. and Bank L.C., (1991) "Creep and Recovery of a Pultruded FRP Frame." *in Advanced Composites Materials in Civil Engineering Structures*, Proceedings of the ASCE Specialty Conference, Las Vegas, Nv, Jan 31-Feb 1, S.L. Iyer and R. Sen. Eds., NY, NY, ASCE, 24-35.

Mosallam, A.S., and Bedewi, N.E., Goldstein, E. (1993) "Structural Behavior of a Universal Connector for PFRP Frame Structures." Submitted for Publication.

Mottram, J., (1991) "Structural Properties of a Pultruded E-Glass Fiber-Reinforced Polymeric I-Beam", University of Warwick, Coventry, U.K.

"Structural Plastics Design Manual," (1984), ASCE manuals and reports on Engineering Practice, No. 63, ASCE, New York, NY.

Werner, R.I. (1984) "Pultrusion Process Engineering-Where Are We Headed?" Proceedings, the 39th Annual Conference, Reinforced Plastics/ Composite Institute, The Society of the Plastic Industry, Inc., January 16-19, Session 1-c, pp 1-8.

Saadatmanesh, H., and Ehsani, M. (1990) "Fiber Composite Plates Can Strengthen Beams", ACI Concrete International: Design Construction, March, 65-71.

Sims, G.D., Johnson, A.F., and Hill, R.D. (1987) "Mechanical and Structural Properties of a GRP Pultruded Section", Composite Structures, 8, 173-187.

Sumerak, J.E. (1985). "Understanding pultrusion process variables for the first time." Plastics Technology, March, 83-85.

Werner, R.I., (1984) "Pultrusion Process Engineering- *Where Are We Headed?*" Proceedings, The 39th Annual Conference, Reinforced Plastics/ Composite Instiute, The Society of the Plastic Industry, Inc., January, 16- 19, Session 1-C, 1-8.

Yen, S, and Williamson, F. (1990) "Accelerated Characterization of Creep Response of an Off-Axis Composite Material." Composite Science and Technology, Vol. 38 , 1990, 119-141.

Zahr, S., Hill, S., Morgan, H., 1993, "Semi-Rigid Behavior of PFRP/UC Beam-To-Column Connections" Proceedings, The ANTEC Conference,(Student Papers Session), Society of Plastic Engineers, New Orleans, Louisiana, May 9-13.

Figure (1) Off-the-Shelf Pultruded Products
*(Bedford Reinforced Plastics, Bedford, PA)*

42                    PLASTICS COMPOSITES

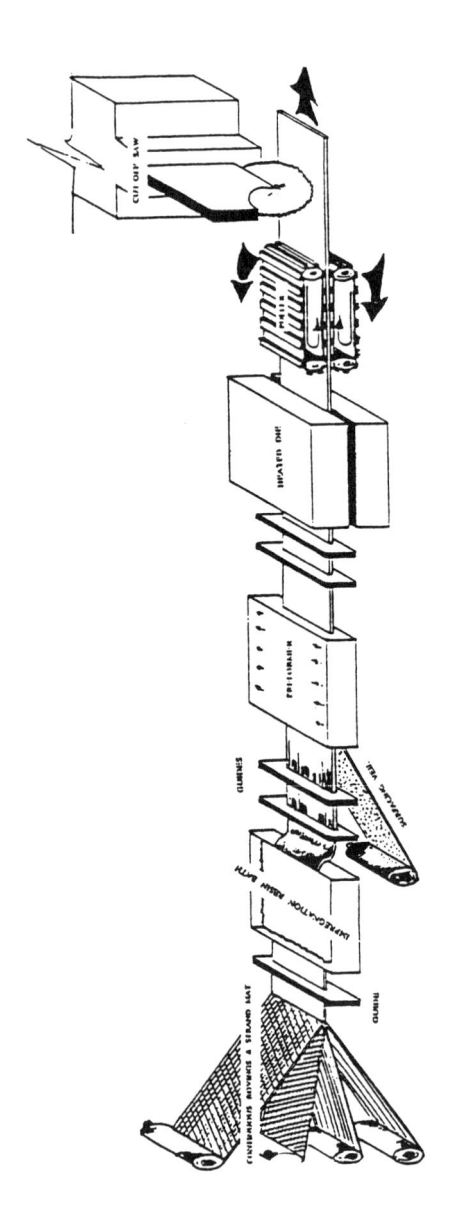

Figure (2) The Steps in the Pultrusion Process
(*Mosallam, 1990*)

Figure (3) A close-up photograph showing the typical web/flange local failure of PFRP open-web *unstiffened* profiles

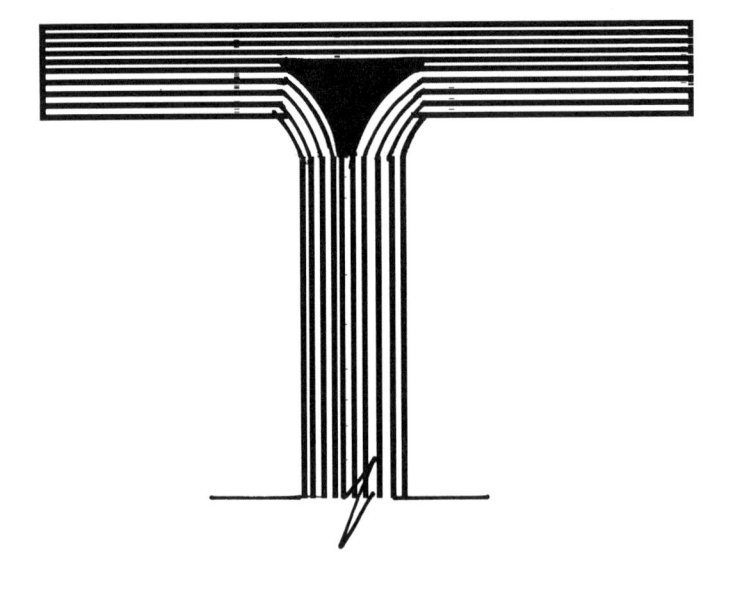

Figure (4) The under-reinforced *inverted triangular* area of open-web PFRP shapes.

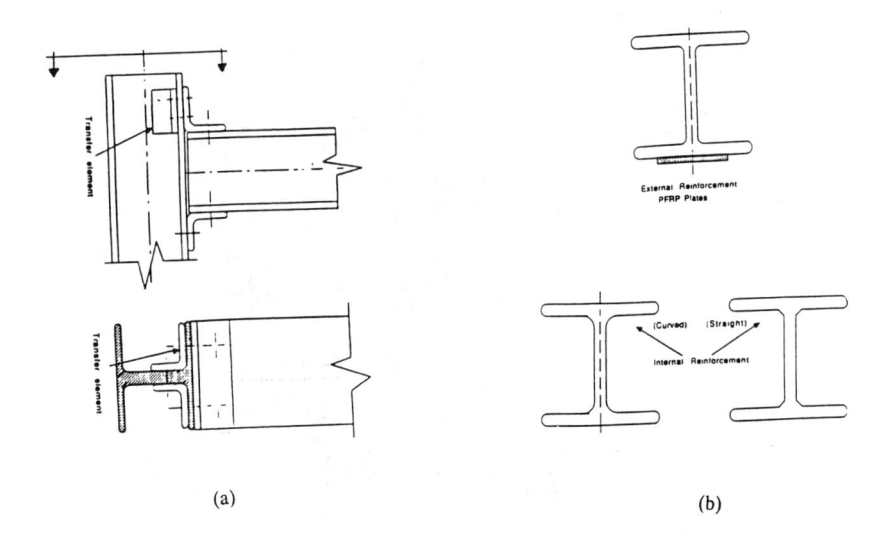

Figure (5)  *a*- PFRP transfer element detail
          *b*- External Reinforcements and suggested details  for web/flange junction (*Mosallam, 1993*)

Figure (6) A common "steel-like" PFRP Frame Connection

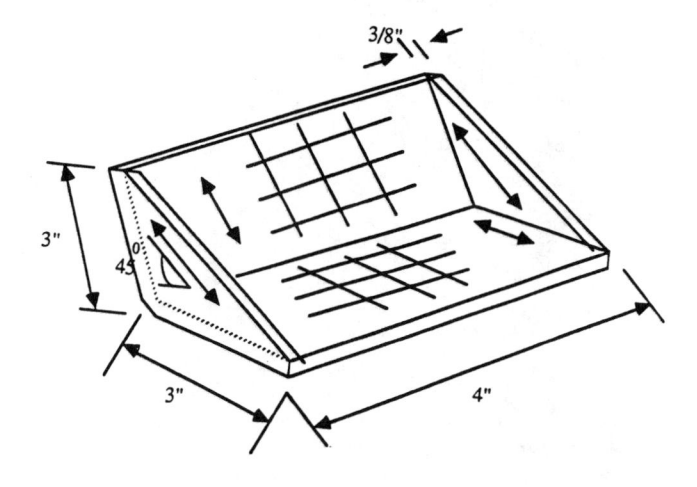

Figure (7) Details of a prototype FRP Universal Connector (UC#4) (*Mosallam and Abdelhamid, 1993*)

Figure (8) Beam-to-Column connection detail using
FRP Universal Connectors (UC)
*(Mosallam, 1993)*

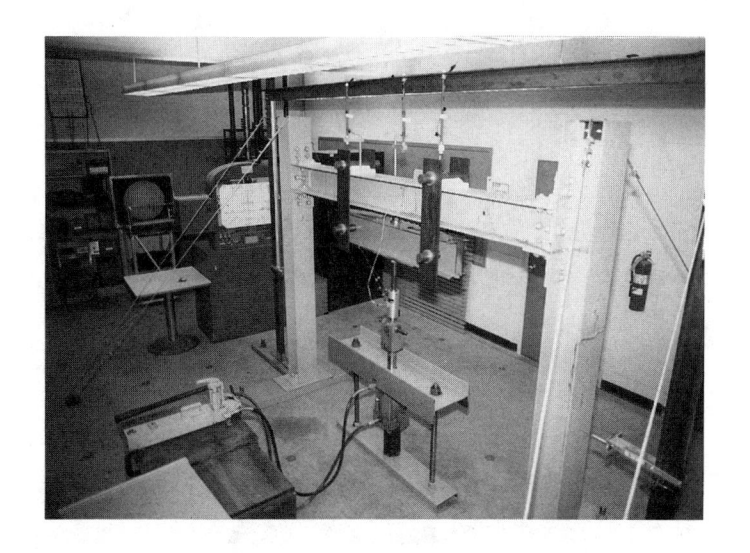

Figure (9) Photograph of a full-scale PFRP portal frame
*(Mosallam, 1993)*

Figure (10) **PFRP** axial coupon creep test setup
*(Mosallam, 1990)*

Figure (11) Test setup of the PFRP dynamic test
*(Mosallam, Abdelhamid, and Conway, 1993)*

Figure (12) Three-story high turrets crown, Orlando, Florida
*(MMFG, Bristol, VA)*

Figure (13) A complete PFRP platform structure
*(IMCO, Inc., Morison Town, NJ)*

Figure (14) PFRP frame structure for EMC testing
*(MMFG, Bristol, VA)*

Figure (15) Complete PFRP frame Structure
*(MMFG, Bristol, VA)*

# 21st CENTURY COMPOSITES REQUIRE ASCE STANDARDS AND SDS

Richard E. Chambers[1], F. ASCE

**Abstract**

ASCE intends to develop consensus standards for structural design fabrication and erection of composites used in civil engineering type structures. This paper describes the ASCE initiative and discusses the philosophy and needs for the structural design system (SDS) of which the standards will be a part. Our perspective is from the standpoint of a practicing consulting materials/structural engineer. Our intent is to acquaint both the practicing civil engineer, and engineers from the composites industry, with this development. Structural products under consideration include structural shapes, and prestressing tendons and reinforcing rods for concrete, and also plastics and composites pipe for other civil engineering applications where standards are not yet available. The effort should enhance the acceptance and proper application of fiber reinforced composites for upgrading our infrastructure in the 21st Century.

## 1. INTRODUCTION

**Purpose**

This paper presents a summary and discussion of the proposed ASCE effort to develop design and construction standards for structural plastics and composites, and the larger effort needed to formalize a comprehensive structural design system for structural applications.

------------------------------
[1] Principal and President, Chambers Engineering P.C., 2356 Washington St., Canton, MA 02021

## Background

Fiber-reinforced plastics (FRP) or composites have been used in a wide range of structural and semi-structural applications in a variety of industries. Fiber-reinforced plastics began to see significant use in the 1950's. The Corvette sports car was introduced in that period; it featured a body manufactured from fiberglass reinforced plastics and still does. Other applications include components of aircraft and aerospace vehicles, boats and buildings, commercial and industrial pipe and tanks for the infrastructure.

## Construction-Related Applications

Corrosion resistance, lower installed cost, light weight and installation ease, and translucence are frequently cited attributes for fiber reinforced composites in construction related applications (ASCE 1984). Perhaps the most significant use in corrosion resistant applications is in pipe and tanks for buried and above-ground use. Some tanks range in size to over 1/2 million gallon capacity and they hold a variety of very aggressive liquids. Fiber-reinforced composites have also been used in building and industrial applications that vary from corrosion resistant ducts, fume hoods, chimneys and walkway gratings, to the framing of crown structures of high-rise buildings, electromagnetically transparent buildings, and structures exposed to corrosive environments. Some impressive shell roof structures were in place by 1962 (Heger 1962), and composite structures were also featured in the 1964-65 New York World's Fair (Heger 1966).

An outstanding example of major use in the infrastructure market is fiberglass reinforced plastic (GFRP) underground petroleum storage tanks (Fig.1). These tanks are frequently the product of choice for use in fuel storage at service stations and other industrial sites. Developed in the 1960's, relatively early in the life of GFRP development, reportedly, over 300,000 such tanks are in now place in the U.S. (Nieshoff 1992).

A more recent example of the use of GFRP in infrastructure is a pedestrian overpass located at the entrance to one of New York City's tunnels (Fig. 2). This follows pioneering work at the University of Virginia where a composites footbridge was constructed and tested in the 1980's (McCormick 1990).

Advanced fiber reinforced plastics (e.g., carbon or graphite and aramid fibers) are relatively more expensive than fiberglass reinforced plastics. In existence for over 25 years, they have seen use in a very diverse range of applications; e.g. sophisticated aerospace structures to golf club shafts.

They are essentially without an experience record in civil engineering construction applications. They will, however, be used to construct a full-scale cable-stayed suspension bridge as a demonstration project at the University of California at San Diego.

Figure 1 -- Underground storage tanks have been a major and successful market for fiberglass reinforced plastics (Photo courtesy of Xerxes Corp.).

Figure 2 -- This recent composites pedestrian overpass at the entrance to a tunnel in New York City demonstrates a trend of increased use in infrastructure (Photo courtesy of MMFG Co.).

## Infrastructure Opportunities

Salt, water, soils and outdoor environments have proven to be more aggressive to conventional structural materials than anticipated by designers of our infrastructure. This has led to a large-scale premature deterioration of pipe, tanks, highways, and bridges and parking structures. This, in turn, has fostered a fresh look at structural composites for infrastructure. Thus, significant efforts are now underway to exploit composites in civil engineering applications (CERF 1993). Applications such as piles, offshore structures and other marine structures, deck structures and suspension cables for bridges, prestressing cables and reinforcing bars and grids for concrete structure, and structural shapes for buildings and bridges are being considered (Faza 1993; Mosallam 1993). In the past, composites have not been incorporated in such applications because of the premium in first costs. As the estimates of the cost of restoring our infrastructure soar however, the life-cycle cost advantage for more durable materials seems to be gaining support and research funding.

As the importance of the structural application grows there is a commensurate increase in the need for appropriate levels of structural reliability. Thus the practicing civil engineer will inevitably become more involved in approving structural plastics and composites for structures that must remain safe and fit for their intended service life. S/he will need quantitative guidelines to evaluate and predict the structural capabilities of new plastics-based products as they come on line. Meanwhile, to facilitate implementation the composites industry and civil engineering community are attempting to understand and respond to the needs of both the engineer who will design, review, specify, and approve such products and the user/owner who should reap the benefits of their use. (Barno 1993; CERF 1993)

## 2. THE ASCE INITIATIVE

The ASCE Structural Plastics Research Council (SPRC), now known as the ASCE Structural Composites and Plastics Committee (SCAP) of the Materials Engineering Division, was formed in the 1960's to advance engineering knowledge and practice in the field of unreinforced and reinforced structural plastics. This group has sought funding and sponsored three major projects to advance engineering of structural plastics, two ASCE Manuals are complete, i.e. the Structural Plastics Design Manual (ASCE 1984) and the Structural Plastics Selection Manual (ASCE 1984), and a report on the Structural Plastics Connections project is near completion. Accordingly, with the accelerated activity and interest in the use of structural plastics and composites in the infrastructure, SCAP formed an ad hoc task group to work with the ASCE Standards Division to obtain approval for ASCE involvement in standards development. The proposal was approved in May of 1993; ASCE is in process of soliciting committee members as of this writing.

The ASCE initiative recognizes the increasing interest of its membership in structural composites in construction. Significant research programs sponsored over the past decade by FHWA and also NSF and ASCE itself, have heightened this interest. It also recognizes that structural plastics are being used in significant projects such as crown structures of high-rise buildings without benefit of authoritative standards and guidelines.

**Scope of Standards Activities**

The ASCE initiative on standards is to bring together consumers, general interest groups, regulatory organizations and producers to develop consensus standards for structural design, fabrication and erection of composites used in civil engineering type structures. The fiber reinforced composites under consideration are typically in the form of structural shapes, prestressing tendons, and reinforcing rods for concrete. The scope of the standards also includes other structural plastics, and certain types of plastics pipe. Currently the broad scope includes standard specifications that cover the following areas:

- Design, fabrication, and erection of composite structural members or buildings and constructed facilities.

- Design and installation of composites and plastics pipe for sewer force main applications, and non-pressure gravity-flow sewer and under drain applications.

The detailed scope of activities and associated priorities awaits development by committee(s) currently being organized by the ASCE Standards Division.

Material standards of ASTM and other appropriate standards organizations will be referenced as standards are developed and become available so there will be no conflict with the standards of other organizations. Where specific needs for modification of ASTM standards emerge from the ASCE effort, liaison with ASTM standards writers will be maintained.

## Products and Applications

The standards for composite structural members preliminarily are expected to address products and applications such as the following:

- "H and I" shaped beams and columns
- Circular, square, and rectangular tubes and plates
- Channels and angles
- Reinforcing rods and cables (coordinated with ACI)
- Gratings and special shapes
- Adhesives
- Rods and threaded rods, nuts

Components of building structures and constructed facilities where the key considerations for selection are attributes such as the following:

- Corrosion resistance
- Electrical resistance
- Electromagnetic radiation transparency
- Weight considerations
- Remedial plate reinforcement

The standards for composites and plastics pipe preliminary will address buried applications for drainage and infrastructure pipe systems where corrosion resistance, lower installed cost, and installation ease are key considerations. Diameters range up to 8 ft. and more for underground use. Significant underground applications for consideration include sewer force main, gravity flow sewer, and drainage pipe. There is anticipation that significant interest will develop for standards on plastic and composite pipe for trench-less ("no-dig") construction projects and for remedial relining of failing existing pipe.

## Priorities

Most likely, the development of the standards for composites structural shapes will be given first priority due to their potential for adoption in the nation's building codes and attendant requirements for public safety, and to reflect the interests expressed by the government, code authorities and representatives of the industry.

The standards for composites and plastics pipe will likely be addressed subsequent to commencement of the standards effort on composites structural shapes and will be coordinated with the Pipeline Division, the Irrigation and Drainage Division, the Urban Subsurface Drainage Standards Committee and other appropriate ASCE committees.

It is anticipated that additional standards for other applications of composites and plastics in construction such as relining of existing pipes, cladding, etc. will be addressed as interest develops after the above priority standards have been initiated.

**Rationale for ASCE Involvement**

No other U.S. standards writings organization has published standards in the areas of design, fabrication, or erection of composite structural members for buildings and constructed facilities nor for design and installation of composites and plastics pipe for the applications under consideration. ASCE is an American National Standards Institute accredited standards writing organization with the capability of promulgating American National Standards which will be nationally recognized and which can be adopted by the model building code groups.

Within its membership ASCE has some 35 members (and growing!) serving on the Structural Composites and Plastics Committee (SCAP). Many of those members have the necessary expertise to develop standards in this field and will serve as the core of standards committee membership. As noted earlier above, the former Structural Plastics Research Council (SCAP's predecessor) developed and published major documents which can be used as resources for standards development. In addition work is underway on the publishing of the report on the state of the art in connections for structural plastics.

**Benefits**

The development of standards and an SDS will lead to very significant benefits, for example:

- Improved reliability and confidence in evaluating adequacy of the installed system for safety and fitness.

- Wider acceptance of new materials and products in engineered applications, as justified by engineering analysis.

- Reduced large scale product tests that are cumbersome, expensive, and necessarily limited in scope.

The major benefits to the public from these standards will accrue in the area of public safety and economy. In addition, this effort will undoubtedly serve to enhance and focus research efforts for use of composites in civil engineering structures for more cost effective research.

We anticipate that public safety will be well served, since the standards on composites structural shapes will likely be adopted in the nation's building codes. The standards will provide guidance to the designer in proper use of these materials and will fill today's void where no authoritative design, fabrication or erection standards are available to address the public safety concern. In addition proper use of the materials will preclude costly over design from lack of knowledge and help to enhance long term performance and service life with lower attendant long-term repair costs, for example, in the infrastructure. Proper use of composites and plastics pipe resulting from the proposed standards will have similar benefits to the public.

## DISCUSSION

The use of structural composites in infrastructure is in the *start-up* mode. That is, to translate the technology and engineering from the expert and proprietary industry-engineered level to the conventional civil engineering level is almost a new venture. Given the start-up mode of the engineering of composites in civil engineering structures, the following is intended to provide a perspective for consideration in the development of standards for composites in construction:

### Characteristics of the Construction Market

While there have been many very successful applications of composites in construction and related structures where structural performance is critical, in others the track record has not been as impressive (Chambers 1992). To a large extent the success or failure is traceable to the level of structural knowledge displayed in the design, manufacture, specification and usage associated with an application or product. Ideally, *total structural knowledge, or TSK,* should be the goal of those involved in the design, manufacture, and construction and use of products where stress, strain or strength are design criteria (Heger 1991). The design and construction standards envisioned by ASCE play a key role in transferring available structural knowledge of a new technology to the construction industry and vice versa, transferring the needs of the construction industry to the composites industry.

The characteristics of the products and the market under consideration herein, are very different than those with which the composites industry has been dealing. Various sectors of the composites industry have been concerned with products that range from space craft to custom-made tanks for storage of aggressive fluids to Corvettes and kayaks - and customers vary from large government funded aerospace companies, to auto makers, to a small business through the small shop of a custom fabricator. The experience base including the standards and specifications development process itself, the technology, and the mode of doing business in general, will of course be different in the construction industry than in any of these markets.

A key factor in understanding the needs of the construction market for composites is that the products under consideration herein are in the domain of the structural engineer; they are an integral part of the structure that s/he designs. The engineer of record has responsibility for the safety and fitness of the infrastructure structure. Since the engineer is the representative of the owner in matters of structure, s/he is in effect the *customer* of the supplier of construction products. S/he is one of thousands working in large and small consulting firms who do most of the infrastructure engineering in this country. The following is a profile of the role of the structural engineer in a design environment, who routinely:

- creates one-off custom-designed-and-built structures for the infrastructure, each valued in the millions to hundreds of millions of dollars,

- synthesizes structures from a myriad of state-of-the-art materials and products,

- works in an extremely cost-competitive environment (usually for low and fixed fees) with no real budget to undertake R & D and testing programs,

- puts together significant structures with a high degree of structural reliability within quantitative limits (prescribed by law in codes),

- endures constant pressure to minimize construction costs, and,

- relies on the guidance of codes and standards, and design aids such as handbooks and computer software to effectively and efficiently deal with a multitude of structural configurations and systems materials and products on a routine basis.

Of course, when employed by public works agencies, the structural engineer may set policy in approving structural products for the infrastructure.

Currently, it is not possible for a practicing civil-structural engineer to follow the above routine and produce designs that incorporate composites as significant structural elements without expending a very significant level of effort beyond the normal scope. One significant obstacle to implementation of composites structural components in infrastructure is the lack of material and product standards. There are no standards for either the fibers and the plastic resin matrix or the products -- structural shapes, reinforcing bars or prestressing rods. Design information is whatever is made available from the manufacturer or can be synthesized from a literature search. The approach to specification and design of structural plastics and composites will have to reflect and accommodate the civil-structural engineer's *modus operandi* in order to facilitate assimilation into infrastructure applications.

In order for composites to be considered as a serious candidate for 21st century construction, both the engineer of record and industry are in need of authoritative guidance for design that has been available for decades for other structural materials (e.g. the AISC Manual for Steel Construction, ACI Code of Practice, NFPA National Design Specifications for Wood Construction) (ACI 1987; AISC 1989; NFPA 1991). This need has been endorsed in various forums by code officials, design professionals, representatives of the U.S. military, and as indicated by the Composites Institute (Barno 1993).

**Structural Design System**

The structural designer depends strongly on a *structural design system*, or *SDS*, for guidance in achieving the goals of design -- safety and fitness -- for the expected life of the (infra) structure (Chambers 1991). That is, an SDS is required in any development involving a significant structural role for composites in constructed facilities and the infrastructure. The SDS is the only practical means by which evaluation of safety and fitness of the finished product can be accomplished.

For our purposes, the SDS reduces TSK discussed above to structural engineering practice. The SDS is a rational set of standardized authoritative rules and protocols governing design, materials, manufacture and fabrication, and construction, that satisfies the goal of design. Structural design systems are in place for all types of structures and structural components for infrastructure and constructed facilities. They are the platform of an engineering practice.

Structural design protocols vary in complexity and sophistication, depending on the risks and the reliability sought for a given type of structure. For example, rules are very detailed for buildings and bridges for the infrastructure where public safety is the dominant concern. Materials and engineering properties are well characterized. Loads and loading cases are well defined. Furthermore, when these rules are referenced in codes, they are law. By contrast, for many structural products the engineer may rely on third party evaluation and approvals -- e.g. the U.L. in the case of GFRP underground storage tanks.

The above types of protocols are sometimes dubbed as ultraconservative provincial barriers to innovation and the development of new construction products, methods and systems. But despite obvious limitations in this regard, existing structural design systems offer the designer enormous flexibility in putting together an impressive array of unique products and structures. Indeed, existing SDS's are not necessarily inflexible as shown by acceptance of the upgrades in the structural performance (and associated standards) of both concrete and steel that have occurred over the years. (Related items such as plastic roofing membranes and plastic sewer, water and DWV pipe have also been accepted to the point where they have captured a significant share of the market from their traditional counterparts.) Unless the rules are changed, and this seems unlikely, an SDS is a given and also necessary if composites are to make inroads in the infrastructure of the 21st Century.

### SDS's for Traditional Materials

SDS's have been in place for traditional structural materials, including wood and concrete, which are arguably more complex and more variable than the compounds used in plastics and composites structural elements. Wood, for example, displays wide variations in strength and stiffness properties. Such factors as loading time, moisture content, checking defects, knots, and angle of the grain all have strong influence in properties. However, with the comprehensive SDS in place, practicing structural engineers deal with these complex materials on a routine basis. They create an impressive variety of structures, readily and with confidence that there will be an acceptable level of success.

### SDS's for Composites

The SDS need not be the same as for the conventional materials noted above. However, the SDS used for both wood and concrete have certain similarities that bear review as background in formulating a system for structural plastics. For example:

- For wood -- the properties of clear, anomaly-free samples of each species are the design basis for strength and modulus of elasticity.

- For concrete -- the compression strength of standard cylinders made from specified materials in specified proportions is the design basis to calculate tension and shear strengths, and also modulus of elasticity.

The designer reflects the effects of the various factors listed above by modifying the basic properties in accordance with prescribed rules. These include effects of loading time on strength and stiffness for both wood and concrete.

There are several sources of approaches to design already in place and accepted for the use of composites in civil engineering-type applications that are available as guidance for the development of full scale SDS for structural composites and plastics. These include but are not limited to the following documents:

- ASCE Structural Plastics Design Manual and Structural Plastics Selection Manual (ASCE 1984; ASCE 1984), and the report on connections also being prepared by ASCE provides a perspective and detailed examples illustrating protocols and concepts appropriate for an SDS.

- ASME codes (ASME 1989) reflect marked advances in standards for the design of glass fiber reinforced plastics used in corrosion-resistant tanks and pressure vessels. Unfortunately, the ASME Codes are silent on materials specifications which detracts from their value.

- ASTM's new design, fabrication and erection standards for structural chimney liners were developed to preclude such major problems encountered at several fossil fuel energy plants in the Midwest (ASTM 1993).

- The venerable MIL HDBK 17 has recently been revised to include the current state-of-the-art in fiber-reinforced-plastic laminates for use in military-type structural application (USDOD 1992)

- The Marine Design Manual is very useful handbook of structural properties available with standardized GFRP laminates of the 1950's. Despite its age, this reference is a useful benchmark for properties of today's fiberglass reinforced plastic laminates (Gibbs & Cox 1960).

- The Building Structural Design Handbook brings together appropriate materials from the ASCE Manual No. 63 as it applies to the structural design of building-type structures (White 1987).

## References

ACI Committee 318, Building Code Requirements for Reiforced Concrete (ACI 318-87), American Concrete Institute, Detroit, MI, 1987.

AISC, *Manual of Steel Construction, 9th ed.*, American Institute of Steel Construction, 1989.

ASCE, *Structural Plastics Design Manual, ASCE Manual of Engineering Practice No. 63*, American Society of Civil Engineers, New York, NY, 1984

ASCE, *Structural Plastics Selection Manual, ASCE Manual of Engineering Practice No. 66*, American Society of Civil Engineers, New York, NY, 1984

ASME, *1989 ASME Boiler and Pressure Vessel Code - Section X, Fiberglass Reinforced Plastic Pressure Vessels*, American Society of Mechanical Engineers, New York, New York, 1989

ASME/ANSI, *RTP-1-1989, Reinforced Plastic Corrosion Resistant Equipment*, American Society of Mechanical Engineers, New York, New York, 1989

ASTM, *D5364 Standard Guide for the Design, Fabrication and Erection of Fiberglass Reinforced Plastics Chimney Liners with Coal-Fired Units*, American Society for Testing and Materials, Philadelphia, PA, 1993.

Barno, Douglas S., *The Use of Structrual Composites in 21st Century Infrastructure Technology*, (proceedings of the Plastics Composites for 21st Century Infrastructure Session, 1993 Annual Convention & Exposition of the ASCE), ASCE, New York, NY, October 1993.

Chambers, R. E., *Composites Performance in the Infrastructure*, (proceedings of Symposium on Composite Materials and Structural Plastics in Civil Engineering Construction), ASCE Materials Engineering Congress, Atlanta, GA, 10-12 Aug. 1992

Chambers, R. E., "Need for a Structural Design System for Composites in Constructed Facilities and the Infrastructure," *Plastics in Building Construction*, Vol. XVI, No. 1, 1991.

Civil Engineering Research Foundation (CERF), Symposium on High Performance - *Materials for Tomorrow's Industry: An Essential Program for America*, Department of Commerce Auditorium, Herbert C. Hoover Building, Washington, DC, April 29, 1993.

Faza, Salem and GangaRao, Hota V. S., *Composites as Future Reinforcements for Concrete Construction*, (proceedings of the Plastics Composites for 21st Century Infrastructure Session, 1993 Annual Convention & Exposition of the ASCE), ASCE, New York, NY, October 1993

Gibbs & Cox, Inc., *Marine Design Manual for Fiberglass Reinforced Plastics*, McGraw-Hill, New York, NY, 1960.

Heger, F. J., "Introduction of New Structural Materials Requires TSK!," *MIT Workshop - Innovative Structures: Materials Design and Construction of the 21st Century*, January 1991

Heger, F. J., Chambers, R. E. and Dietz, A. G. H., *Design Analysis and Economics of Fiberglass Reinforced Plastic World's Fair Structures*, (proceedings of the 21st Annual Reinforced Plastics Technical & Management Conference), Society of the Plastics Industry Inc., New York, 1966

Heger, F. J., Chambers, R. E., and Dietz, A. G. H., "On the Use of Plastics and Other Composite Materials for Shell Roof Structures," (proceedings of the World Conference on Shell Structures), San Francisco, CA, October 1962

McCormick, Fred C., *Laboratory Fatigue Investigation of a GRP Bridge, Serviceability and Durability of Construction Materials*, (proceedings of the First Materials Engineering Congress), ASCE, New York, NY, August 1990.

Mosallam, Ayman S., *Pultruded Composites: Materials for the 21st Century*, (proceedings of the Plastics Composites for 21st Century Infrastructure Session, 1993 Annual Convention & Exposition of the ASCE), ASCE, New York, NY, October 1993

National Forest Products Association (NFPA), *National Design Specifications for Wood Construction*, NDS, NFPA, Washington, DC, 1991. (See also supplement: *Design Values for Wood Construction*, NFPA, 1991.)

Nieshoff, E. C., *Private Communication*, Fiberglass Petroleum Tank and Pipe Institute, May 1992

USDOD, *Polymer Matrix Composites, MIL HDBK 17, 1C & 3G, Guidelines, Utilization of Data*, US Department of Defense, Washington, DC, 1992

White, Richard N. and Salmon, Charles G., *Building Structural Design Handbook*, John Wiley & Sons, Inc., New York, NY, 1987

## SUBJECT INDEX
### Page number refers to first page of paper

Codes, 56
Composite materials, 1, 15, 23
Composite structures, 1, 15, 56
Concrete, reinforced, 1, 15, 56
Creep, 23

Dynamic response, 23

Fiber reinforced materials, 15
Fiber reinforced plastics, 1, 15, 23, 56

Fiberglass, 1

Infrastructure, 1, 23, 56

Performance evaluation, 23

Standards, 56
Structural design, 56
Structure reinforcement, 15

## AUTHOR INDEX
### Page number refers to first page of paper

Barno, Douglas S., 1

Chambers, Richard E., 56

Faza, Salem S., 15

GangaRao, Hota V. S., 15

Mosallam, Ayman S., 23